THE NEW NATURALIST
A SURVEY OF BRITISH NATURAL HISTORY

FLOWERS OF THE COAST

The aim of this series is to interest the general reader in the wild life of Britain by recapturing the inquiring spirit of the old naturalists. The Editors believe that the natural pride of the British public in the native fauna and flora, to which must be added concern for their conservation, is best fostered by maintaining a high standard of accuracy combined with clarity of exposition in presenting the results of modern scientific research. The plants and animals are described in relation to their homes and habitats and are portrayed in the full beauty of their natural colours, by the latest methods of colour photography and reproduction.

THE NEW NATURALIST

FLOWERS OF THE COAST

by

IAN HEPBURN

WITH A CHAPTER
ON COASTAL PHYSIOGRAPHY
BY J. A. STEERS

WITH 17 COLOUR PHOTOGRAPHS
BY JOHN MARKHAM AND OTHERS
43 BLACK AND WHITE PHOTOGRAPHS
AND 14 LINE DRAWINGS AND MAPS

COLLINS
ST JAMES'S PLACE
LONDON

to

PHYLLIS

*who loves the sea but is sometimes
uncertain of her botany*

ISBN 0 00 213067 X

*First published 1952
Reprinted 1962
Reprinted 1966
Reprinted 1972
Printed in Great Britain
Collins Clear-Type Press : London and Glasgow*

CONTENTS

CONTENTS

COLOUR PLATES

It should be noted that throughout this book Plate numbers in arabic figures refer to Colour Plates, while roman numerals are used for Black and White Plates

vii

PLATES IN BLACK AND WHITE

ix

EDITORS' PREFACE

In any civilised country the two types of habitat least altered by man
or his domesticated animals are the mountains and the coast. *Mountains
and Moorlands* have already been dealt with in this series by Professor
W. H. Pearsall, and, in the present volume, Mr. Ian Hepburn tackles
the wild flowers of our coast. Despite the modest disclaimer in his
Preface, Mr. Hepburn is particularly well qualified for the task. He
is Second Master at Oundle School and one of the leading amateur
plant ecologists in the country. He has served on the Council of the
British Ecological Society for many years and has published the results
of a number of his ecological investigations, including work on the
plants of the Northamptonshire limestone and of sea-cliffs. As one
would expect, his approach to coastal vegetation is primarily an
ecological one, but he has by no means neglected the beauty or
botanical interest of the plants themselves, and we feel that he gives
in his book a balanced and vivid account of a fascinating subject.

Coastal vegetation has always had a particular attraction for field
botanists, and the reasons are not difficult to understand. First, in no
other habitat does the flora change so completely in the course of a
short walk; one minute there is sand filling our shoes, and marram-
grass pricking our legs, and the next a level carpet of sea-lavender lies
around us, the dunes are left behind and a grey cliff, with samphire
and tree-mallow, lures us farther on. In Britain, probably more than
any other country, is this ever-varying quality of the coastline manifest;
and the maritime vegetation shows a corresponding richness. Secondly,
many of the plants themselves are of great beauty and, in some cases,
rarity. The lovely glaucous leaves of the oyster-plant setting off its
blue and pink flowers against a Scottish shingle bank, or the autumn
squill sprinkling a dry September cliff-turf in Cornwall, spring to the
mind—and there are many others. Thirdly, for an ecologist, the coast
has a special interest, as nowhere are the succession of vegetation and
its dependence on a varying environment so evident, and nowhere

have they been more thoroughly investigated. As Mr. Hepburn shows, in an acre or two of salt-marsh are displayed, for those who have eyes to see, most of the more important principles of stable vegetational succession, while on a newly fallen section of cliff, the gradual colonisation of virgin ground can be studied year after year. Lastly, if the environment for the plants is unrivalled in its variety, it is no less so for the humans who study them; a botanist who chooses the British coasts as his hunting ground can enjoy, with a clear conscience, a succession of seaside holidays on what is perhaps the finest coastline in the world.

THE EDITORS

AUTHOR'S PREFACE

IT WAS with considerable diffidence that I, a mere amateur, accepted the invitation of the Editors to write this book. Nor has the subsequent appearance of other books in this series, all written by experts, done anything to dispel this feeling. Nevertheless, I am grateful to them for being so insistent that I should try my hand, for I have greatly enjoyed doing it, and certainly know rather more about the subject than I did when I started!

Ever since I began to take an interest in wild flowers I have always found maritime plants especially attractive. I was lucky enough to go to school within five miles of Blakeney, and first learnt the common seaside plants on the Cley marshes, the Weybourne shingle bank, and sometimes on the historic Blakeney Point itself. Later, I had the good fortune to live for a number of years in north Cornwall, where, in addition to fine stretches of cliffs, excellent sand-dunes and small areas of salt-marsh were within easy reach. Coastal vegetation, therefore, is very much in my blood, and if I have been able, in this book, to convey something of the pleasure I have had in botanising along the coast, I shall be very well satisfied. Now that it is written, my first wish is to get out to the coast once more to start some field-work. Perhaps I may express the hope that others may be similarly stimulated.

I am deeply indebted to many authors, whose books and original papers I have freely consulted. A full list of these, with acknowledgments, appears in the Bibliography. I count myself very fortunate in having persuaded Professor Steers to write the chapter on the physiography of the coast. No one else possesses his detailed knowledge of the whole coastline of Britain, and he should really have written the whole book himself, for he is no mean botanist. I am specially grateful to John Markham for taking so much trouble in obtaining many of the photographs I required—whatever may be said about the text, it will be generally admitted that his pictures are first-class.

Finally my grateful thanks are due to John Gilmour for his continual encouragement and advice at all stages.

For permission to reproduce figures appearing in the text acknowledgment is made to the following sources:

Cambridge University Press: Figs. 6, 9c, and 11, from *The British Islands and their Vegetation* (1939) by A. G. Tansley.

G. Bell and Sons, Limited: Figs. 4, 10, and 12 from *Plant Form and Function* (1946) by F. E. Fritsch and E. J. Salisbury.

Royal Geographical Society: Fig. 1 from *The Geographical Journal* (1937).

CHAPTER I

INTRODUCTORY

THERE IS NO NEED to possess any great knowledge of botany to see at once that a number of plants grow round our coasts which are never seen inland. It matters little whether you explore a salt-marsh, a sand-dune area, a shingly beach, or some rocky cliffs; in any of these places you can count on coming across some, at any rate, of these purely coastal plants. Moreover, where the conditions are suitable, they often occur in large numbers and are chiefly responsible for the great difference in appearance between seaside and inland vegetation as a whole. In point of fact the characteristic look of the vegetation is often due to the presence of comparatively few coastal species, for along most of the coastline many familiar inland plants can be seen as well. Indeed it is only in salt-marshes, where the whole area is regularly washed by the tides, that practically the whole plant population is composed of maritime species. Most of those growing on sand-dunes, for instance, are also found inland but, since the highly distinctive marram-grass is the dominant species in nearly all areas of blown sand, the vegetation of the whole area acquires its characteristic " coastal " appearance.

It is no good pretending that all these seaside plants are things of beauty. Many of them possess thick fleshy leaves, which do not give them a graceful form, and they often have untidy woody stems. They are apt to have a rather weather-beaten appearance, which is hardly surprising when one remembers the conditions to which they are habitually exposed. No one could suggest that the prickly saltwort (*Salsola kali*) (Pl. I, p. 6), the sea-beet (*Beta maritima*) (Pl. XXXVIII, p. 199) or the annual glasswort (*Salicornia stricta*) (Pl. XIII, p. 114) are attractive to the eye, though like most coastal plants they look interesting because they are unfamiliar. Such charm as they possess is largely associated with the exciting sort of places where they can be found. Yet there are some maritime plants which are as beautiful as any in the land, as the colour illustrations in this book show, and when seen together in a mass produce some glorious colour effects. There are few sights to compare with a West Country cliff in May, when thousands of sea-pinks (*Armeria maritima*) and sea-campions (*Silene maritima*) are in full bloom together in their natural rock garden against a background of deep blue sea (Pl. II, p. 7), or with the startling brilliance of a large sheet of sea-lavender (*Limonium vulgare*) on some East Anglian salt-marsh in late summer (Pl. 5, p. 87). In a quieter way, too, the long waving leaves of marram-grass (*Ammophila arenaria*) on the crests of shifting sand-dunes have an irresistible charm of their own at all times of the year (Pl. III, p. 14).

Coastal vegetation exhibits many features of particular interest to botanists. Perhaps its greatest attraction lies in the fact that it has, as a whole, been comparatively little altered by man. In a small area like the British Isles there are few places where the clearing of natural forests, the ploughing up of the land, or heavy grazing by domestic animals has not modified or completely transformed the natural vegetation. The thin strip of land bordering the coastline has, however, largely escaped these attentions, since for a number of reasons it does not lend itself easily to agricultural development. Even so, arable farming is frequently carried on quite close to the edges of cliffs, and the higher levels of most salt-marshes are grazed by cattle and sheep. Yet it is not too much to say that practically anywhere along the coast, except in the vicinity of a seaside resort or a port, one can count on finding some patches of maritime vegetation approximately in its natural state. It may be a belt of sand-dunes at the top of a small beach, a little area of salt-marsh around the mouth of a river, or just a collection of foreshore plants on the sand or shingle of the shore, yet they reproduce on a small scale many of the characteristic features of more extensive areas of the same kind of vegetation elsewhere.

Although the coastline has suffered relatively little at the hands of farmers, it is always open to a more serious menace. A justifiable desire of many people to spend their holidays by the sea has led to a great deal of ill-considered building during recent years, in many cases preventing public access to the shore or cliffs. Much irreparable harm has already been done, but the passage of the National Parks Act in 1949 and the setting up of the Nature Conservancy is encouraging evidence that the public has at last woken up to the danger of progressive spoliation of the countryside, and it is to be hoped that a halt may eventually be called to any further defacement. As things are at present, there are fortunately long stretches of unspoilt coast still available for botanical study in many parts of the country, though one could weep over the damage which has been done in the past along some of the more populated parts of the shore.

The coastline provides a number of extremely varied types of habitat in which plants may grow. Salt-marshes, sand-dunes, rocky cliffs and shingle beaches, to mention a few of them, all differ widely from each other in the demands they make upon plants, and each is found to bear a type of vegetation peculiar to its own special conditions. As we shall have occasion to employ the word "habitat" many times in this book it may be as well to make clear at once that it does not merely refer to the place in which a plant is found growing, but includes all those conditions influencing its growth which are connected with the place. In point of fact the main conditions remain remarkably constant in each of the various coastal habitats all over Britain and, as a result, the general type of vegetation found in each of them is strikingly similar wherever it occurs, although there are naturally many local differences of detail.

This similarity can best be appreciated if the botanist, when observing coastal plants, pays special attention to those growing together most frequently in a particular habitat, instead of merely searching for individual (and perhaps rare) species. If he does this, it will soon become apparent to him that many of them habitually occur in well-marked and easily recognised communities. Incidentally, the plant communities found along the coast are, for the most part, simpler in composition than those found inland. This makes them ideal material for study by anyone who is getting beyond the stage of just collecting and identifying plants (a very necessary stage for all field-botanists), and is beginning to take an interest in the vegetation as a whole.

Let it, however, be said at once, for the benefit of the pure plant-hunter, that the coastal regions do in fact provide happy hunting grounds for many relatively rare plants, largely because they have been so little disturbed. This is particularly true of sand-dune areas, which often carry a rich flora, although the rarities found there are usually not confined to the coastal belt. At the same time, I feel strongly that many naturalists would enjoy their botanising even more if they cultivated the habit of looking at the whole vegetation as a unit, noticing how the individual species of which it is composed are grouped, and trying to discover why particular plant communities grow where they do. After all, by doing this, one is not denied the excitement of coming across an unusual plant, and it has the great advantage that the common plants become once more interesting, since their relative frequency is of fundamental importance in determining the nature of the general vegetation. The relation of vegetation to its surroundings is known as plant ecology, and nowhere can a start at this way of botanising be better made than along the coast. In the hope that some of my readers may be stimulated to look at seaside plants in this broader way, the characteristic vegetation of the main coastal habitats is discussed from a general ecological standpoint in the later chapters of this book. With this end in view, I have devoted Chapter 3 to a brief explanation of some of the main principles of plant ecology.

It is important to remember that the coastline never remains completely static, but is continually in process of being either eroded or built up. The ways in which this can take place are discussed in some detail in the next chapter, but it is clear that both erosion and deposition can provide virgin ground, suitable for colonisation by plants. One of the most interesting features of coastal vegetation is that it is often possible to observe a whole series of successive stages whereby these bare habitats acquire a comparatively stable plant-cover, from the first tentative seedlings to such complex communities as are typical of heathland or scrub. This fascinating process in which one community is replaced by another, and this in turn by others, can be witnessed and understood far better along the coast than anywhere else. Unfortunately nowhere is it possible to observe the production under natural conditions of the final or " climax " stage (see page 29) in the development of the vegetation.

But perhaps the best reason of all for studying the flowers of the coast is that by so-doing one visits delightful and exciting places, always within sight and sound of the sea. For my own part, I frankly admit that this was why I became specially interested in seaside plants. Although I have botanised in all sorts of country, and in particular greatly enjoy doing so amongst mountains, I have never experienced anywhere the same variety and pleasure as I have had when looking for flowers along the coast.

A complete list of all the plants growing round the coast of the British Isles would be an extremely long one, but the great majority would be species found equally frequently in quite different habitats far away from the sea. In particular, sand-dunes, cliff-tops, and shingle that is no longer exposed to wave-action, can all support a wide variety of inland plants. It is, however, with those plants which are confined to the coastal belt that we are chiefly concerned in this book, and these fall into two main groups, only the first of which consists of genuine maritime species. This includes practically all the plants found in regularly submerged salt-marshes, and also a number of others occurring in habitats which are commonly exposed to a certain amount of sea-spray, such as beaches and cliffs. The plants belonging to this group are called *halophytes* (Greek *halos* =sea salt), and may be roughly described as plants which can live in soils where the water is salt. The main difference between these and normal plants is that their protoplasm (living material) is not destroyed by exposure, either externally or internally, to salt solutions.

One is tempted to divide this group further into " true halophytes," chiefly confined to salt-marshes and thriving under conditions where the actual water-table remains permanently saline, and what may be called " spray halophytes," characteristic of habitats exposed to salt spray, but not normally submerged by sea water. Annual seablite (*Suaeda maritima*) (Pl. 3, p. 71) and common sea-lavender (*Limonium vulgare*) (Pl. 5, p. 87) are examples of true halophytes, while spray halophytes are exemplified by samphire (*Crithmum maritimum*) (Pl. 10, p. 142) or sea-rocket (*Cakile maritima*) (Pl. XVI, p. 83). We hardly know enough yet about the requirements of many of these plants, however, to make such a distinction and it is therefore inadvisable to carry it far. Some true halophytes, for instance, appear actually to require the presence of salt in their root-water for their full development, but probably all spray halophytes and certainly some " true " ones (e.g. thrift) can grow equally well in ordinary garden soil, provided that competition from taller and faster-growing plants is excluded. Indeed the absence of competition from other plants, which are unable to endure the conditions, is the chief reason for the abundance of spray halophytes on exposed cliffs and similar habitats. Since the term " halophyte " is used to describe plants of such differing requirements, it is probably best to consider that the one essential feature they have in common is that they can *tolerate* a certain amount of salt water, rather than that they actually *require* its presence. Moreover, it should not be forgotten that sea-water contains small quantities of many other salts besides sodium chloride, and it is quite possible that these may also have important effects on the growth of plants, which we do not at present understand. The majority of halophytes are perennials and often develop extensive woody rootstocks. But their most obvious physical feature is usually their fleshy or succulent leaves, a characteristic which appears to be closely connected with the absorption of salt water, since it is also shown by a number of inland plants when they grow near the sea or inland salt areas.

The second group of plants confined to the coastal belt consists of those found only on sand-dunes. There is, of course, nothing inherently maritime about a sand-dune, but it so happens that in the British Isles the main dune areas are coastal. The largest areas of blown sand in the world are found in desert regions far from the sea, and in North America extensive " coastal " dunes exist round the Great Lakes, where the water is fresh. Even in this country considerable patches of blown sand, carrying some typical dune plants, can be seen in the Breckland. All plants capable of establishing themselves in mobile sand bear a general resemblance to each other wherever they occur, and the majority of them are known as *xerophytes* (Greek *xeros* =dry). Most of them possess unusually extensive roots or rhizomes (buried stems), and their leaves are often adapted in some way or other to cut down excessive loss of water from them. Many xerophytes have succulent leaves and stems, similar to those belonging to halophytes. This likeness, is, however, largely accidental and is due in this case to the development of water-holding cells, called collectively " aqueous tissue " (see page 47), for the purpose of storing water. Xerophytes are by no means confined to dune areas in Britain but may also be found in other dry habitats inland. Those confined to dunes, besides possessing an unusual ability to withstand drought, are also able to deal with the problem of occasional burial in loose sand by their exceptional powers of sending up new shoots as soon as they are covered over—marram-grass (Fig. 10, p. 51) possessing this power to a quite remarkable degree. Other familiar examples of dune-xerophytes are sea-holly (*Eryngium maritimum*) (Pl. 1, p. 35) and sea-sandwort (*Honckenya peploides*). Shingle beaches and dry cliffs are other coastal habitats where plants of this group commonly occur. The characteristic form and habit adopted by both

Plate I Prickly saltwort, *Salsola kali*; a typical strand plant of sandy shores, Norfolk.

John Markham

Plate II Sea-pinks, *Armeria maritima*, in a natural rock-garden. Lundy
Cove, Cornwall.

halophytes and xerophytes is discussed in greater detail in Chapter 4.
The remainder of the plants to be seen in the coastal belt are too numerous and varied for it to be worth while attempting further classification. Practically all types of inland vegetation are represented somewhere along the coast, although woodland is virtually absent from all exposed situations and is confined to the borders of sheltered estuaries or to deep valleys (or coombes) which have been cut through cliffs, where sufficient protection from the wind is provided. Typical marsh plants, for instance, grow in the wet " slacks " sometimes found between ranges of sand-dunes (Pl. XXII, p. 111), and in north-west Scotland and Ireland, alpine plants, characteristic of the higher mountains in Britain, occur near sea-level on the cliffs and beaches just as they do farther north in sub-arctic regions. Nor is it surprising that a large number of common plants, usually associated with dry waste-places of all kinds, find a home on the older stabilised dunes and shingle beaches, particularly shallow-rooted annuals, which can pass through their life-cycle during the winter and spring, before the advent of the summer droughts.

There are, however, a number of what may be called " submaritime " plants, which are neither halophytes nor xerophytes, but seem rarely to be found more than a few miles from the sea. Well-known examples of these are the slender thistle (*Carduus tenuiflorus*), alexanders (*Smyrnium olusatrum*) (Pl. IV, p. 15), fennel (*Foeniculum vulgare*) (Pl. XXXIV, p. 175), and one of the mouse-ear chickweeds (*Cerastium tetrandrum*). Besides these, there is a much longer list of species which, though not confined to the coastal belt, are always far commoner there than farther inland. Storksbill (*Erodium cicutarium*) and buck's-horn plantain (*Plantago coronopus*) (Pl. XXXVI, p. 179) are familiar examples. None of these plants conform to any special type (there are a number of aquatic plants as well) and no explanation is at present forthcoming to account for their distribution (see p. 160).

It is unavoidable in a book of this kind that it should contain a large number of lists of the plants usually found in the various habitats. These may well seem tedious to some readers who are unfamiliar with the appearance of many of the species mentioned, and I have therefore devoted Chapter 12 to giving brief popular descriptions of plants mainly confined to the coastal belt. I have also added notes on their distribution and their relative importance in the general vegetation. It should be clearly understood, however, that it is in no sense my object to provide a flora for identifying all the plants likely to be encountered along the coast, and I have not attempted to include those species which are equally common inland. A number of standard floras are mentioned in the Bibliography (p. 226), and one of these should be consulted when identifications are being made. At the same time, it is to be hoped that this chapter may prove useful for quick reference. The main object of this book, however, is to describe seaside vegetation as a whole, and to relate it as far as possible to the various habitats where it is found.

I have devoted a separate chapter to the vegetation found in each of the main habitats. The following summary will give some idea of particular portions of the coast with which we shall be concerned in each chapter:

1. SALT-MARSHES *(Chapter 5)*
   The mud or sand of the inter-tidal zone on flat shores, which is sufficiently protected from violent wave-action to support vegetation, including areas only flooded by the highest tides.

2. BEACHES AND FORESHORES *(Chapter 6)*
   A comparatively narrow zone along the tops of exposed beaches of any material, other than rocks, only reached by the highest tides.

3. SAND-DUNES. *(Chapter 7)*
   All areas of sand, originating in material blown by the wind from the shore, from shifting open dunes just beginning to accumulate round individual plants, to old and mature deposits whose surface has become almost completely stabilised by a close cover of vegetation.

4. SHINGLE BEACHES *(Chapter 8)*
   The beaches, bars or spits of water-worn pebbles, derived from rocks by wave-erosion and deposited by the sea on low-lying shores. These vary from highly unstable banks of stones, continually shifted by the waves, to the oldest beaches where the shingle is completely dormant and has long been isolated from any wave-action.

5. CLIFFS AND ROCKS *(Chapter 9)*
   Rocky places above the high-tide mark or cliffs of any material,

which are to some extent exposed to salt spray. Certain artificial habitats such as sea-walls, which are similarly exposed to spray, are included in this category.

6. CLIFF-TOPS (*Chapter* 10)

The strip of ground along the tops of cliffs, exposed to a certain amount of spray and often supporting some characteristic sub-maritime plants as well as maritime and inland species. When the tops are level this is usually quite narrow, but on steep slopes it may cover a large area and then differs from the cliff habitats in the previous chapter only in possessing more soil.

7. BRACKISH WATER (*Chapter* 11)

The swamps, ditches, lagoons, or estuaries of slow-moving rivers where the water remains brackish as a result of sea-water mixing with fresh water. These are often inhabited by a characteristic mixture of sub-maritime and inland aquatic plants.

It is hardly necessary to add that these habitats are by no means always sharply separated, but in many places overlap considerably. Sand-dunes, for instance, often grow up on the crests of shingle ridges, and salt-marshes develop easily behind the protection of shingle or sand-bars. As a result, the vegetation is often mixed up in a distinctly confusing manner—nowhere better seen than along the north Norfolk coast.

CHAPTER 2

THE PHYSIOGRAPHICAL BACKGROUND
A Résumé

by

J. A. STEERS

THIS BOOK is concerned with the ecology of the sea-coast and the seashore. The various types of ground that come under this broad title are subject not only to constant change but often to violent change. Even the hardest cliffs are comparatively unstable, and almost always subject to strong winds and storm-waves which may do much superficial damage, even if the actual rate at which the cliff retreats is, in terms of human life, extremely slow.

Great Britain has a remarkably long and intricate coastline and a long and varied geological history. Strata of nearly every period are well represented. These rocks, and the associated igneous rocks (also of very different ages) give the coast great interest and variety. We can observe the white and often perpendicular cliffs of the Chalk, the magnificent ranges of dark red cliffs of Old Red Sandstone in Caithness and Kincardine, the grey walls of Carboniferous Limestone with their flat grassy tops in west Pembrokeshire, the rapidly wasting cliffs of glacial deposits of north Norfolk and Holderness, the heavily glaciated cliffs of the whitish-grey Lewisian Gneiss, alternating with those of the brown and often spectacular Torridon Sandstone in north-western Scotland, and the granite cliffs of Land's End and the Isles of Scilly. These are but a few examples from many; the point is that the rock type alone—quite apart from whether the beds are folded, broken, horizontal, or cut into diverse forms by marine or sub-aerial erosion —makes the coastline extremely interesting.

Cliffs

In the preceding paragraphs the word cliff has been used but not defined. The Oxford English Dictionary gives the following meanings for the word:

1. A perpendicular or steep face of rock of considerable height, usually implying that the strata are broken and exposed in section; an escarpment.
2. (especially in modern use) A perpendicular face of rock on the seashore, or (less usually) overhanging a lake or river.

These definitions are descriptive, comprehensive, and do not pre-suppose any particular origin. Only too often it is assumed that because cliffs face the sea, they are wholly the result of marine erosion. The mere presence of plants on a cliff face is not, of course, evidence that no erosion is taking place; it may suggest that erosion is fairly slow, or that it may operate by large and infrequent slips. But a steep, grassy slope running down to near sea-level may be quite untouched by marine erosion. Some of the so-called Hog's Back cliffs of north Devon are only affected by the sea in their lowest parts; their upper slopes have been produced in some other way (Pl. V, p. 22).

Many cliffs are found at the back of a flat or platform where they are no longer washed by the waves. Along the Durham coast this is so, and still more along many miles of the coast of Scotland, especially in Galloway, Kintyre, Arran, Ayrshire, and many other districts. Sea-level, relative to the land, has altered since the platform was formed, and where the alteration is considerable, the old cliff may now be well away from the sea and its vegetation only indirectly influenced by it. The boulder-beach at the cliff-foot south of Duncansby Head in Caithness indicates a slight change of level, since the boulders are lichen-covered and are somewhat above the normal height which wave attack reaches. In the Gower peninsula, the limestone cliffs run right down to sea-level, often in unbroken slopes. But in many small inlets traces of raised beaches are found, and it follows that the outer cliffs, even if they do run down to and below sea-level, cannot wholly be the product of modern marine erosion. In parts of Scotland there are sometimes two or three old beaches found in the same place, often with a normal shingle flora. Between the mouth of the Findhorn and Burghead (Moray Firth) there is a great series of ridges, all well above the present sea-level. To-day the sea is cutting into them and making a shingle cliff fifteen or more feet high. Their natural vegetation closely resembles that of modern beaches, but has undergone changes as a result of its more stable position. In many other places around the Moray Firth, especially between Hilton of Cadboll and Tarbat Ness, and in various localities on the west coast, particularly on Islay, Colonsay, and Jura, magnificent expanses of shingle-flats and old cliffs are to be seen.

In other places the sea-level has risen relative to the land, so that the lower valleys are flooded, and what were cliffs are now submarine slopes. Much of the beauty of Pembrokeshire, Cornwall and Devon depends on the influx of the tides into long and intricate inlets like Milford Haven, Plymouth Sound, the Camel estuary and the mouth of the Dart. These were carved first by rivers and streams on the land, later drowned by a relative rise of sea-level. Only their outer cliffs have been modified by marine action; the sheltered inlets now often contain salt-marshes. The pulsation of the tides into their innermost recesses, and the intimate relation between land and sea vegetation create beauty.

In western Scotland there is another type of inlet, the sea-loch or fiord. Many of the narrow straits between islands or between the mainland and adjacent islands are similar. These lochs usually owe their present appearance to ice action which has scoured out, widened, and perhaps deepened and straightened pre-existing river valleys, which sometimes followed lines of weakness produced originally by faulting. The sides of these lochs and straits are cliffs in the sense of the dictionary definition, but are not the product of marine erosion. Up the fiords there are often salt-marshes, and at low water the expanse of golden, brown, red and varicoloured seaweeds, surrounded by the higher marsh plants, under a strong sun is unforgettable.

Cliff-form is usually closely related to structure. If the beds are more or less horizontal and thick only one may form the cliff, but more often two or more drop out. At Hunstanton the brown Carstone overlain by the Red Chalk and this again by the White Chalk makes a spectacular cliff, subject to rapid erosion because of the ease with which the sea eats into the soft Carstone, thus producing falls. Near Lyme Regis, and in Glamorgan, the nearly vertical cliffs are composed of thin beds of limestone and clays or shales arranged horizontally. Elsewhere, the beds may be steeply inclined seawards or landwards, folded or faulted, and the form of the cliff will depend much on the trend of the cliffs relative to that of the folds. The rocks of the Isle of Wight and the Isle of Purbeck are strongly folded in an east-west direction. The cliffs along their south coasts run with the beds which, seen from the sea, appear almost horizontal. But if one sails round Durlston Point to Poole Harbour, or between Ventnor and Ryde, the beds are seen on end, and the way in which the sea has cut into the softer ones is evident. Where folding is acute and intricate and the rocks are hard, the sculpture is often bizarre. A view from a boat close inshore between Boscastle and Hartland Point, or between Berwick and Cockburnspath, or around the west of Pembrokeshire reveals details of surprising interest.

Where all the rocks are soft and perhaps geologically young, changes go on at a quicker rate than where the rocks are harder and older. Changes are rapid between Flamborough Head (Pl. XXVI, p. 131) and the Isle of Wight. If the Angles and Saxons could revisit the country they certainly would not recognise it; on the other hand, the Phoenicians might see little difference in the Cornish coast, except near St. Michael's Mount. Even now there is good reason to suppose a slow sinking of south-eastern England. Since the end of the Ice Age there have been great changes in the levels of land and sea, and these have often had a more profound effect on the present appearance of the coast than have erosion and accretion.

BEACHES

If the beach along a fairly straight coast is examined, it will usually be noticed that if shingle is present it is at or near the top. Sand and finer particles occur lower down, and usually the fineness of material increases seawards. The cliffs behind the shingle may be of any kind, and are not necessarily the source of the shingle. After a severe storm much or all of the beach material can be removed, and the platform on which it rests exposed. In the succeeding normal weather, the beach will gradually accumulate again. Even after an ordinary blow the beach may be combed down, so that coarse and fine material are much more mixed.

When waves break, beach material, coarse and fine, is churned up. There is often some order and arrangement in this movement. If the waves are approaching the shore at right-angles, the pebbles and small

stones move up and down the beach. The waves break and send up
the beach sheets of water called the swash or send, which carry material
upwards. Some of the water of the swash percolates into the beach,
some returns to the sea as the backwash. This is nearly always less
powerful than the swash, but in its deeper parts can move a good deal
of material. If, however, the waves approach the beach obliquely,
so also may they advance up it, and stones and sand are not merely
carried upwards, but also sideways. When the swash dies out, the
backwash returns directly down the slope, and any material moved
by it travels in the same direction. Thus on open coasts on to which
the waves come obliquely, there is a great deal of lateral displacement
of beach material. This process is called beach-drifting, and is of the
utmost importance. Its effect is often seen where groynes or break-
waters are built athwart the beach to hold material travelling along it
by this process. The beach on one side of a groyne is usually higher
than on the other, although often after a storm from a different quarter
the high and low sides may temporarily change places.[1]

The waves also have a sorting effect, and drive the stones to higher
parts of the beach. This process can often be seen in action while
waves are breaking on a beach of mixed material. There is still
another important factor. Around our coasts there is usually a noticeable difference in the level of the water at low and high tide. On the
open coast the range seldom exceeds twenty feet, but in bays and gulfs
it may be more. The highest rises and the lowest falls occur about
the times of new and full moon. At the half-moons the difference
between high and low water is small. Suppose it is now the moon's
first quarter, and the weather generally fair. The waves at both the
morning and evening tide will reach to much about the same level.
But in succeeding days the high waters will rise higher and the low
waters will fall lower with each succeeding tide until the time of full
moon.[2]   What effect will this have on the beach? If marked pebbles
have been scattered near the water line at the time of the moon's
first quarter they will be seen not only to have moved along the beach
if the waves are oblique, but to have been pushed up it by waves at each

[1]The beach, whether it is natural or accumulated by the help of groynes, is the
natural protection to the coast, since the waves expend their energy mainly on it.
Beach-drifting by oblique waves usually implies a beach with a reasonable slope.
On a very flat beach it may be negligible.
[2]For reasons that need not be discussed here, there is usually a delay of 24 hours
or so, so that the highest tides may be a day or two after Full Moon.

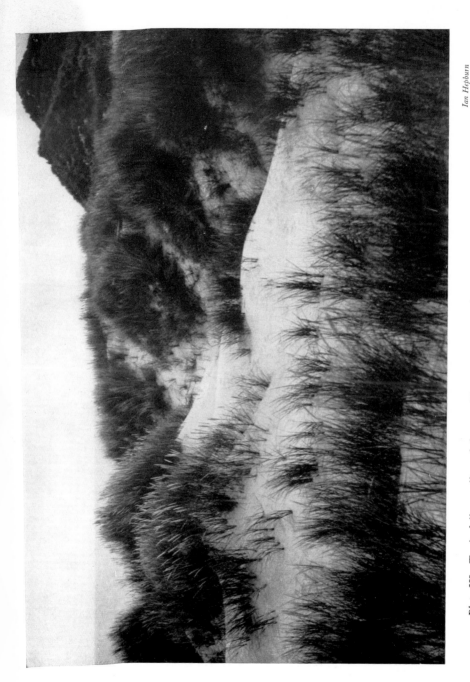

Ian Hepburn

Plate III Typical " open " sand-dunes with marram-grass, *Ammophila arenaria*, in flower. More mature " fixed " dunes in background (right). Scolt Head, Norfolk.

John Markham

Plate IV Alexanders, *Smyrnium olusatrum*; a typical submaritime plant
of waysides near the sea. Norfolk.

successive tide, and have gathered near the top, usually in an existing ridge. During the subsequent fall of tide-level after full moon, the pebbles are left stranded, and they may only just be reached again at the next period of springs at the time of new moon.

Thus, if nothing else happened, the pebbles might remain at the beach top for ever. But two other factors are likely to affect them. First, the swing of the tides from neaps to springs is only part of a larger swing that shows itself in particularly high tides near the equinoxes, and sometimes at other times of the year. Secondly, if a severe storm attacks a coast, especially at a period of big tides, shingle may be either swept far above its normal level or dragged in large quantities down the beach. An exceptional storm may overtop the highest beaches. The effect of ordinary storms is plain along any shingle beach, since the seaward face is frequently marked by minor ridges parallel to its length. These are either the heights reached by the last high tides, or the limits of recent storms.

Quite apart from these up-and-down and lateral movements of some beach material, vast quantities of finer stuff are moved alongshore by a different process. A bather on a sandy shelving beach in ordinary weather and in water three or four feet deep notices the lifting effect of the waves: the sand is at the same time somewhat disturbed about his feet. If he allows himself to float off such a beach, he notices that the tidal current carries him one way or another along it. The sand stirred up by the waves may, when a tidal current is running with any speed, also be carried sideways for a short distance. Picture this process during a tidal cycle, and during rough and stormy conditions, and it is at once apparent what vast quantities of fine material can be carried along a beach.

Current-action, however, only takes place under water and below the zone of wave-break. It operates on the higher parts of the beach only at or near high tide; on the lower parts, on an open coast the current may run one way at or near low water, and the opposite way at about high water. Two things at least follow—first, in the deeper water, material may move in different ways at different stages of the tide, and whether there is a balance of movement will depend upon the relative strengths of the flood and ebb currents. Secondly, on the parts of the beach covered only at high water, the movement of material is likely to be in one direction only—that of the current at the time of high water. The resultant process is called long-shore drifting. With beach-drifting it is of the utmost importance in the study of shoreline phenomena.

A wave breaks when it enters water the depth of which is approximately half its wave-length. Thus on a shallow coast, big waves break farther out than do small ones. When breaking offshore, the waves—just as on a beach—drive material up in front of them, so that sometimes ridges of sand and shingle are built some distance from the original shore. If a shingle ridge of this sort attains a fair degree of stability, it becomes an outer beach along which beach-drifting can take place. Hence the ridge may lengthen and become what is called an offshore bar (see page 120). If the process continues a lagoon-like expanse may be enclosed between it and the old shoreline. Offshore bars are seldom unbroken for long distances, since there are often gaps through which the tide enters and leaves the lagoon, in which marsh development is favoured.

If the supply of shingle is great, and if the lateral transport along a coast is marked, the shingle can accumulate in great forelands like Orfordness, Dungeness, the Crumbles, and the shingle ridges off the Culbin Sands and other parts of the Moray Firth shore. At an early stage a ridge is built. After a time the new shingle coming along the coast shallows the sea floor off the first ridge, so that the waves build another in front of it. This may go on until a whole series of such ridges is formed. It is often noticed that at one part of a shingle foreland the ridges run out to sea in such a way that it is clear they are suffering erosion, whereas at another part new shingle is accreting and being built up into ridges. A study of any big shingle foreland will illustrate this process, but there are few more striking examples than the shingle formation known as the Bar, near Nairn. Fig. 1 shows that it is composed of a number of individual ridges, the north-eastern ends of which are being eroded, whereas growth is continuous at the other end. In short, the whole structure is slowly shifting along the coast.

Along the south-west facing side of Dungeness there are many ridges running directly out to sea, and obviously at one time they continued for some distance. Erosion has cut them, and the material thus provided has travelled round the point of Dungeness and gradually helped to build the numerous ridges forming Denge Beach. Erosion is constantly taking place on the one side, accretion on the other.

The shingle that composes the banks and beaches comes partly

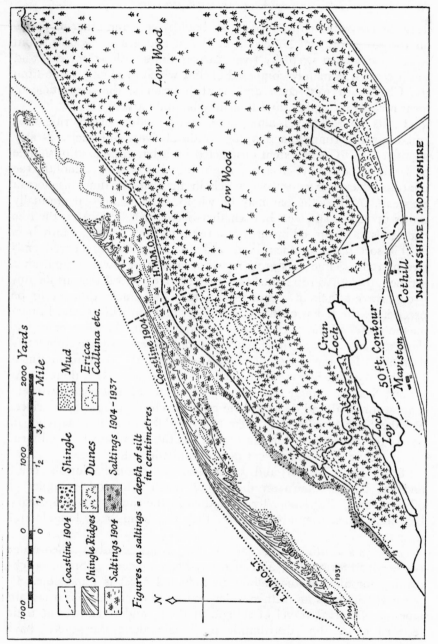

Legend (on map):

1000 2000 Yards

1000 1000

0 1/4 1/2 3/4 1 Mile

Coastline 1904

Shingle Ridges

Saltings 1904

Shingle

Dunes

Saltings 1904–1937

Mud

Erica Calluna etc.

Figures on saltings = depth of silt in centimetres

N

Map labels: Low Wood, Low Wood, H.W.M.O.S.T, Coastline 1904, Cran Loch, Loch Loy, 50 ft. Contour, Maviston, Cothill, NAIRNSHIRE | MORAYSHIRE, L.W.M.O.S.T, 1937, 1904

Fig. 1.—The Bar (from Steers, *Geogr. Journal*, 1937).

from the erosion of cliffs, partly from boulder clay and other materials on the sea floor, and largely from glacial and gravel deposits, from which it has been swept by rivers in past times. Along our east and south coast it is mainly composed of flint which originally came from the Chalk. In west coast and Scottish beaches, the percentage of local rocks is far higher, and flint may be quite absent.

In its lateral travel alongshore, shingle often builds ridges or embankments, running across the mouths of rivers and inlets. Nearly all rivers are to some extent obstructed by a bar composed of shingle or sand, or both. When a bar is growing across a river mouth, the unattached end extends very much in the same manner as does a tip heap. At first the bar may be wholly below water; it gradually grows up to the surface. But whether above or below water, the free end tends to be turned inwards as a result of wave action. Many bars of this sort have on their landward side laterals or recurved ends (see p. 120). Some bars grow forward, later turn inwards, and, after a time, grow forward again. It is easy to give general reasons for this —e.g. wave-attack in a storm—but it is extremely difficult to be precise. If the bar is obstructing a river, its form will depend in part on the power of the river to keep its mouth clear. Small streams like those at Chideock and other places on the Dorset coast are completely dammed. In others, e.g. the Exe and Teign, the river maintains a mouth; the Exe Bar is particularly interesting since it is double. At Orford Ness the shingle has not only formed a bar, but has grown into a great foreland and deflected the river for eleven miles. Some rivers like the Spey usually keep their mouths through shingle beaches in nearly the same place, whereas others, like the neighbouring Findhorn, by no means free from violent floods, are deflected.

Scolt Head Island and other features of the Norfolk coast and the Bar off Nairn on the Moray Firth are good examples of offshore bars, and, they, too, lengthen in the same way as ordinary beaches. Since they are offshore, they can send back long lateral ridges.

In bays and other inlets, shingle beaches are usually washed up at the head, to form what are called bay-head beaches. Normal beach-drifting for any distance along an indented coast is impossible. If, however, the bays are rather wider and more open, there is a certain amount of lateral travel of shingle in them, and it gathers at their leeward ends. The distribution of shingle along the several bays between Pwllheli and Penrhyndeudraeth is most instructive. Sometimes a ridge forms across the mid-part of a bay, such as Cemlyn Bay in Anglesey.

DUNES

Shingle ridges of various kinds have been discussed at some length because they form the " skeleton " combining the " flesh " of dunes and salt-marsh. By no means all dunes are built on shingle ridges, and many ridges have no dunes. But if a shingle ridge is being formed in a locality where an expanse of sand is bared at low tide, the wind blowing over the sand will carry much of it on to the ridge and deposit some of it in its interior and some on its surface. This may cause dune growth. But (see Chapter 7) the real dune-builders are the sea couch-grass (*Agropyron junceiforme*), marram-grass (*Ammophila arenaria*), and sea lyme-grass (*Elymus arenarius*) that take root in the shingle, send up shoots, and so begin to trap the sand. Marram-grass is remarkable in this way, thriving best where the sand supply is most prolific (Pl. III, p. 14). If another shingle ridge is built in front of the old one, the same process will begin on that, and the dunes on the inner ridge will be, at any rate partially, deprived of their sand supply. The dunes may remain, but are often partly blown away, whereas those on the newer ridge increase. Under favourable circumstances they grow and become permanent features. In England dunes seldom exceed fifty or sixty feet in height if built up from sea-level.

The evolution of coastal dunes depends largely on the available sand supply, the vegetation, the prevalent winds, and the nature of the foundation on which they rest. With a constant wind and sand supply they may attain large sizes, but the vegetation rarely forms a close covering, and there is always the chance of minor hollows being enlarged. This is especially the case in older dunes, somewhat cut off from their sand supply. Often these are riddled with blow-outs (Pl. XXI, p. 110) so that mere fragments of the original sand-dune chain remain. On some dunes the wind may enlarge a hollow, but new sand may sometimes replace that blown away.

Groups of coastal dunes are often arranged in lines, despite their superficial irregularity, and the lines correspond with the trends of old shingle ridges. This is shown at Blakeney Point, Scolt Head Island, and Morfa Harlech. But on Morfa Harlech and even more on Morfa

C

Dyffryn this relatively simple arrangement is often confused by later movement. In both places there are numerous blow-outs, and the inner dunes have locally been set in motion again, so that they are advancing over the low ground on their landward side. On the other hand, at Culbin the major dunes covered groups of ridges and showed no relationship to their trends. The same may be true at Forvie, but unfortunately no detail is known of the surface upon which the dunes rest. Sometimes a line of coastal dunes advances downwind more quickly in its middle and higher parts, than near its lower ends. This is because it is easier for the plants to hold the ends than it is for them to restrain the sand on the higher parts once it has begun to blow again. A dune-line that has advanced in this particular way, when well-developed, is called a parabolic dune. The best examples in Great Britain are at Maviston (really a part of the Culbin Sands) where the dunes have advanced over and buried well-grown forest trees (Pl. VI, p. 23).

Every possible transition can be found around our dune shores from the tiny heap of sand just beginning to gather round a tuft of grass, or a small obstruction on a beach, through the elementary stage of a foredune based on a shingle ridge, to the more evolved forms on the older ridges which have not been deprived of their sand supply. If the sand supply is great and it blows along or oblique to the shingle ridges, rather than at right angles to them, there is no particular reason for the dunes to have any correlation in trend with the ridges, and they may grow to considerable heights. When the dunes have no relation to the ridges it is possible that there has been a change of direction of the prevalent winds. The old shingle ridges at Culbin are part of a raised beach system; the dunes are the result of conditions extending back only two or three centuries.

When dunes are deprived of a sand supply they take on a dead appearance. The wind may regain mastery, and most of the sand may be blown away, leaving hummocks behind, known as remanié dunes. If the sand blown away begins to accumulate, as well it may, quite near at hand, the observer may be confused by the contiguity of old and new dunes; the new ones in this case may sometimes be to landward of the old. What is more, the new dunes will almost certainly have a covering of fresh-looking marram-grass, although the sand is derived from the old ones! It is essential to keep in mind this liability to rapid change in dunes; a return to a familiar place after an interval of five or ten years may easily result in the visitor being temporarily lost! Sometimes the sand blown by strong winds is spread out as a layer over considerable areas. To some extent this happens in any dunes; the sandy pasture inside Morfa Harlech is an instance. In the Western Isles of Scotland there is a special development known as the machair (see p. 114). True, it is often associated with prominent lines of dune, but the machair proper usually lies landward of dunes and shingle bars, and forms, as in South Uist and Tiree where it is particularly well developed, a low sandy plain between sea and peat and hills. It is primarily shell-sand and provides a most fertile sward. In high summer the numberless plants and flowers, the strong scent, the vivid colours, the blue sea and white beach, and the wide open views and great expanse of sky make the machair the centre of a most beautiful landscape.

SALT-MARSHES

Salt-marshes and their growth are described in detail in Chapter 5. They show better than any other feature the intimate relation between ecology and physiography in coastal evolution. They are built up of the mud, silt, and fine sand carried by the tidal (and other) currents, and deposited in quiet places. Their nature will depend much on the material of which they are composed; there is a great difference (see pp. 84 and 86) between the firm mud marshes of Norfolk and the sandy wastes of Morecambe Bay. Marsh growth begins in certain favoured places on the sea floor, especially on that part of it which is bared at low tide. Ideal conditions obtain on the coast of North Norfolk where formations like Blakeney Point, Wells Headland, Scolt Head Island, and the small high-tide island at Thornham provide shelter. Marsh growth takes place outwards from the original coast and inwards from the protecting ridge. Moreover, where structures like Blakeney Point throw off a number of laterals, or recurved ends, the best possible conditions for marsh formation often exist between neighbouring laterals, especially if they are so built as to make a narrow entrance to a relatively broad expanse inside (e.g. the Marram marshes at Blakeney Point). Into these quiet backwaters tidal water pours, and stands quietly for a period before the ebb (Pl. XIa, p. 62). Sedimentation takes place readily, and since the water has to drain out through the narrow mouths, material deposited round the margins is undisturbed. Hence, a considerable quantity of mud soon gathers. The same thing happens in less enclosed places, but the speed of accumulation is generally less.[1] Between Wells and Blakeney there are extensive marshes, only here and there fringed on their seaward side by a ridge. The larger waves break well out, so that in the quiet inshore water sedimentation can proceed.

The same quiet conditions often prevail in parts of estuaries and other embayments. There is considerable marsh growth in Hamford Water, around Canvey Island and Sheppey in the Thames Estuary, in Chichester Harbour, in Southampton Water and Poole Harbour, on the upper parts of Plymouth Sound and Milford Haven, and in the Bristol Channel. Many other places round our coasts show similar growth. The principles of accumulation are similar everywhere and need not be analysed in further detail.

Reference has been made to beach-drifting, to the movement of dunes, and to tidal and other currents. Let us look generally at Great Britain in relation to wind-systems. The prevalent winds (i.e. those blowing most frequently) in any part of this country are from a westerly direction, usually somewhat south-westerly in England and Wales. On our western shores the dominant winds (i.e. those having greatest power or effect) also blow from the same general direction. On the east coast, however, the dominant winds come from the quarter between north and east. How any wind will affect a particular stretch of beach must depend greatly on local conditions. To take an example: south-westerly winds will have great effect in Mount's Bay, but not just east of the Lizard Peninsula. Allowing, however, for detail of this kind, the westerlies are responsible in the main for eastward directed beach-drift along the Channel, for that up the Bristol Channel, and for that along the coasts of Cardigan Bay. Another factor is also important—the relation of wind-direction to the amount of open water off a particular coast. On the Cumberland coast, the direction of beach-drift is north and south from approximately St. Bee's Head. This is in general conformity with the amount of open water off these two parts of the coast. However, the relationship is better seen on the

[1]Since the mud is exposed at low water, and in many places for a period of several tides (i.e. where only spring tides reach) the mud is deposited in thin layers. Often sand from dunes is blown on to marshes, so that layers of sand and mud alternate. Near the seaward part of a marsh, shingle also may be thrown over it as a result of a storm breaking through or washing over the protecting ridge.

Plate V Hog's-back cliffs near Combe Martin, North Devon, from
the air.

Plate VI Remains of a pine forest, overwhelmed by advancing sand and later laid bare by a further shift of sand. Maviston (Culbin) dunes, Moray Firth.

east side of England. Along the Norfolk coast, excluding minor exceptions, the travel of beach material is on the whole westwards from Sheringham along the north coast and south-east and south from that same place along the east coast of the county. The dominant winds and waves approaching the Cromer-Sheringham coast are from the quarter between north and east: these, working in with the extent of open water offshore and with the general trend of the coast, are mainly responsible for the outward drift from that locality. The southward drift of beach material continues, apart from a few minor interruptions, as far as the Thames.

On the whole (except in the inner parts of the Firths of Tay and Forth) beach material travels southwards from north-eastern Aberdeenshire all down the east coast of Great Britain. It is, however, along the more open coast south of Flamborough Head that this is most noticeable. Along the south shore of the Moray Firth and the coast as far as Banff and even Rosehearty the general movement of beach material is to the west, and southwards from Wick it is also directed towards Dornoch Firth and Inverness.

On an indented coast of hard rocks it is difficult to generalise. Each separate bay usually has its own beach, and whatever solid stuff travels round the enclosing headlands does so below water level and cannot easily be traced. The individual coves of Cornwall, Devon, Pembrokeshire, the north coast of Scotland, and elsewhere may have their beaches temporarily removed by storms, but they will gather again in normal times. It is probably true to say that each bay has its own shingle and sand economy. On relatively deep water coasts, such as that of the west of Scotland, it is impossible to generalise about the travel of sand and silt.

The main contrasts we have made between the different parts of the coasts of Great Britain may perhaps be related to an even more general factor. Apart from the Lancashire coast, and excluding local occurrences of boulder clay, it is approximately true to say that a line joining the mouth of the Exe to that of the Tees separates a region of softer rocks and simpler structure to the south and east from a more complicated region of harder rocks to the west and north. The former is associated with long lines of open beach and sweeping curves along which beach and longshore drifting are well exemplified. The latter is often a coast broken by inlets and hard and rocky lines of cliff, along which lateral movement is irregular.

CHAPTER 3

SOME ECOLOGICAL
CONSIDERATIONS

IT IS HARDLY possible to understand how the vegetation is distributed round the coast-line without having some slight acquaintance with the principles of plant ecology. In this chapter we shall therefore consider quite briefly what ecology is about and also take the opportunity to explain some of the terms which are commonly used by ecologists. No attempt will be made to go more deeply into the subject than is necessary to follow the method used in the later chapters, which describe in detail the characteristic vegetation to be found in various typical habitats along the shore. For a fuller account of the subject the reader is referred to Professor A. G. Tansley's fine book, *The British Islands and their Vegetation*. In the following short account the examples have been chosen as far as possible from seaside vegetation in the hope that the main characteristics of coastal habitats in general will become apparent.

Plant ecology is concerned with the study of plants in their natural habitats and their relations with their surroundings. It is thus primarily a field study and can be worked out only in the place where the plants are actually growing. The present popularity of both plant and animal ecology is to a certain extent a reaction from some of the more specialised lines of inquiry in biology, which have to be carried out indoors in laboratories.

One of the most fundamental differences between plants and animals is that the former are fixed in the soil, and cannot therefore move about when they are growing. They are thus, of necessity, gregarious and have to lead a communal life. Plants are, in fact, usually found in well-marked communities, whose composition depends

24

on the nature of the habitat and a number of other factors, some of which are discussed later in this chapter. The word *plant community* is a general one which is used to describe any collection of plants growing together which can be said to possess a definite individuality. If there is much bare ground between the individual plants, which is available for colonisation by other species, the community is said to be *open*. The plants found growing on the front (seaward) range of sand-dunes in an area of blown sand form a typical open community (Pl. VII, p. 30). Other obvious examples to be found amongst coastal vegetation are the communities inhabiting exposed sea-cliffs (Pl. II, p. 7), and the mobile mud along the edges of salt-marshes (Pl. XIII, p. 66). When the vegetation is more or less continuous, and competition for the available space becomes an important factor, the community is said to be *closed*. An open community generally represents an early stage in the colonisation of an area, but it may also be found in a habitat where the conditions are so harsh that plants have great difficulty in existing at all.

Although the individual members of an open community depend largely on the nature of the habitat, the amounts and nature of the species present will depend increasingly on their inter-relations in the available space. Usually one or more *dominant species*, which are mainly responsible for the general appearance of the community, can be recognised. They are frequently the tallest-growing plants present and may thus exercise a profound influence upon the other inhabitants of the community, particularly by competing successfully for the available space or by causing shade. As examples from coastal vegetation, we may mention rice-grass (*Spartina townsendii*), which is the main dominant species in the communities formed on the soft mud of salt-marshes along the south coast (Pl. XIV, p. 67), and the sea-rush (*Juncus maritimus*), which frequently dominates a zone along the upper edges of salt-marshes elsewhere. The other species associated with these dominants are known as *subordinate* species. If these are found in nearly every example of a community, they are called *constant* species. Any other plants which turn up from time to time in the community, but are not really characteristic, are known as *casuals*.

Plant communities may be of very different sizes and importance, and it is customary to divide them into various classes. The largest unit of vegetation is called a *plant formation* and usually refers to a broad type of vegetation which remains roughly the same over a whole continent or even throughout the world. The character of a formation depends on the nature of the habitat and it reflects this in the distinctive life-forms of its principal species. Thus the Salt-marsh Formation contains a highly characteristic population of halophytes, whose specialised life-forms reflect the most important feature of the habitat, that of its periodical immersion by sea-water. Similarly the Sand-dune Formation contains another very characteristic population of plants, many of which are xerophytes and specially adapted to grow in the semi-arid conditions of blown sand. (Some ecologists restrict the use of this term to the ultimate climax vegetation which can be developed in a habitat under given climatic conditions, and would not therefore refer to either of these essentially transitional types as formations.)

The term *plant association* has in the past been used to refer to so many different units of vegetation that, to avoid confusion, it has not been employed in this book. It is now generally accepted that it should be used to describe a relatively large unit, usually a geographical sub-division of a formation which is characterised by a particular dominant species. As an example, we could say that the Oak-Beech Association is the typical form in the British Isles of the main European Deciduous Forest Formation. In the same way, the Marram-grass Association is typical of the Sand-dune Formation in this country, though associations with other plants as dominants may be found in similar habitats in other parts of the world.

From the point of view of our discussion of coastal vegetation, however, the most important unit to define is the *plant consociation*. This is a smaller affair than either of those so far mentioned, although it was frequently called an association in the old days. It consists of a community with (usually) a single dominant species. Salt-marshes generally show well-marked examples, since the vegetation often occurs in distinct zones. Thus the lowest strip is often dominated by annual glasswort (*Salicornia stricta*) (Pl. XIII, p. 66), and other typical zones are dominated by such plants as sea-aster (*Aster tripolium*) sea manna-grass (*Puccinellia maritima*) (Pl. XIb, p. 62), sea-lavender (*Limonium vulgare*) (Pl. 5, p. 87), etc. Plant consociations are often named after the Latin name of their dominant species by adding the suffix *etum* to the stem of the Latin name of the genus. Thus the consociations referred to above are usually called the *Salicornietum, Asteretum, Puccinellietum* and *Limonietum* respectively. Should there be any possibility of confusion over the identity of the dominant species, the specific name is usually added in the genitive case. For example, consociations dominated by two separate rushes are found in salt-marshes in different areas, and the word *Juncetum maritimae* is therefore used for that dominated by the sea-rush, to distinguish it from that dominated by the mud-rush, which is called *Juncetum gerardii*.

The smallest unit with which we need concern ourselves is the *plant society*. This is a purely local community, which may sometimes be noticed within a consociation, dominated by a species which would be considered a subordinate one if the consociation were viewed as a whole. Societies generally owe their origin to some small local differences in the habitat. Thus the sea-purslane (*Halimione* (*Obione*) *portulacoides*) often forms a distinct society along the sides of the creeks which cut through the *Puccinellietum* or *Asteretum* in a salt-marsh, because the soil there is better drained (Pl. XV, p. 82). Another type of society is a *layer society*, which can be observed when the vegetation is composed of plants of very different heights. This is most obvious in a forest, but an important society of mosses and lichens can often be seen below the main herbaceous layer on the older sand-dunes, and there is frequently a layer of shade-loving plants in the *Juncetum maritimae* in a salt-marsh.

In explaining the various units of vegetation which are recognised by ecologists we have tacitly assumed that they remain stable and possess a constant composition and structure. This is, however, by no means the case; nearly all vegetation is continually changing, although the rate at which this is proceeding varies greatly. Some communities appear to be remarkably stable, but others are mere passing phases, which soon give place to others. We ought therefore to look upon all these units as representing positions of relative equilibrium into which plants group themselves for a time. Generally speaking, the changes which are in progress all tend towards a position of greater stability. All progressive change of this kind is known as *succession*.

There are two main types of change which can bring about a succession of vegetation. To the first type belong all those which are caused by purely physical factors which alter the habitat in some way, making it less suitable for the first occupants and more suitable for others. A long-term example of this kind of change would be a gradual alteration in the climate; there is plenty of geological evidence of the effect of such climatic changes in past eras upon the vegetation of the

British Isles. It is often possible, however, to see much more rapid changes in progress. For example, the sand on the sea-shore always contains enough salt to make it somewhat alkaline, but as soon as it has been raised above the level of the highest tides in the form of a sand-dune, the salt will rapidly be washed out by the rain. If there is only a small amount of calcium carbonate (another substance causing alkalinity) in the sand, this will also in time be washed out from the surface layers, and plants which prefer more acid conditions can then become established. In this way, the first colonists, which prefer neutral or slightly limy soils, will be gradually replaced by others and eventually " dune-heath," with heather as the dominant species, may sometimes be produced. Another example is provided in some dune areas, where water tends to accumulate between the ranges of the older dunes, producing a totally different type of habitat within the main area of blown sand. Here a community consisting almost entirely of marsh plants frequently appears. Yet another example of the effect of a physical change can often be seen in salt-marshes, where the tide has been artificially excluded from the upper levels by the construction of some sort of barrier. Here the vegetation is rapidly changed by the appearance of numbers of non-halophytes as soon as the rain has washed out the residual salt from the soil.

The second type of change which may bring about a succession of vegetation is one produced by the plants themselves. When any bare ground is colonised, there is nearly always at first a fairly rapid series of changes in the composition of the plant communities. The first colonists or *pioneers* will almost invariably give way to others later, and these in turn may afterwards be replaced by still others until a relatively stable equilibrium is reached between the habitat and its vegetation. This type of development can probably be observed taking place along the coast better than anywhere else in the country. The usual way in which plants alter a habitat is by adding humus to it. Humus is the dark organic material produced by the partial decay of plant remains, and as the first colonists die off this material begins to accumulate in the surface layers. In course of time the physical properties of the soil are modified by the addition of this humus and, in particular, its water-holding power is steadily increased. As a result of this, it becomes possible for a wider selection of plants to gain a footing. As a rule, the new occupants are of greater size and stronger growth, so that the earlier colonists are eventually swamped by them. Later on, these in turn may be choked out by other even stronger plants. Thus each successive community, by modifying the soil, tends to make the habitat more suitable for the growth of new species, but in so doing lays the way open for its own ultimate destruction. For example, many of the early colonists in the mobile sand of young sand-dunes are unable to exist in the thick sward of grasses and other plants which cover the surface of the older dunes, and the marram-grass itself is eventually stifled when the surface of the sand becomes completely fixed. The early colonists of sand-dunes, however, not only add humus to the sand but also modify the habitat by anchoring the surface of the sand. Only a limited number of pioneer plants can exist in the shifting sand between the clumps of marram-grass on the young dunes, but they all make their contribution towards fixing the surface of the sand. As a result of their efforts, it gradually becomes possible for a greater variety of plants to become established, and eventually the characteristic close sward of fixed dunes is produced.

In many cases a modification of the habitat may be produced by the combined efforts of plants and physical factors. The colonisation of the bare soft mud on the edge of a salt-marsh is a good example of this. The pioneer plants, such as glasswort or rice-grass, are instrumental in stabilising the mud and also add humus to it. In addition, they aid the natural physical process in which mud is deposited by causing a distinct slackening of the tide as it ebbs and flows over them, and in this way the level of the habitat is gradually raised and stabilised so that other plants can become established.

Generally speaking, all succession is directed towards developing the most complex vegetation which the climate will permit, no matter what the nature of the original habitat may have been. The ultimate vegetation produced in this way is called the *climax formation* or the *climatic climax*. The communities making up this formation will be more or less stable and will not be seriously threatened by new invaders. In most of England and the southern part of Scotland, if the vegetation were left completely undisturbed, oak or beech forest would eventually be developed. In the north of Scotland and most of the central portion also, if we exclude the tops of the higher mountains, the climatic climax would, however, be pine forest, an association of the Northern Coniferous Forest Formation. In comparatively recent times, most of the British Isles was forested in this way, but the large-scale felling of our woodlands during the Middle Ages and later has almost obliterated the natural forests. Nowadays, as a result of intensive agricultural operations, the climax formation is rarely reached in the course of natural succession. Where suitable areas exist, which are not cultivated or grazed, the absence of suitable seed-parents in the immediate neighbourhood precludes the development of natural woodland. Ecologists recognise, however, a number of relatively stable *sub-climaxes* in the vegetation of this country, which are developed under the conditions which are normally present.

Any natural succession of communities which replace each other in a particular habitat is called a *sere*. Thus those which succeed each other in a salt-marsh all belong to the *halosere*, salt being the master-factor controlling each of them, and those developing on blown sand to the *psammosere* (Greek: *psammos* =sand). The sea-coast provides practically the only habitats in this country where one can see a more or less complete series of communities starting with bare ground and finishing with a type of vegetation which remains comparatively stable under the particular conditions. Elsewhere, succession can be most easily observed in an area which has previously carried some fairly stable type of vegetation, but which has subsequently been modified in some way or other. This is well illustrated when a wood is felled or a heath is burnt and is known as *secondary succession*. Good examples of this type of development can also be seen along the coast, as for instance when the surface vegetation on a sand-dune is broken through and the strong winds produce a " blow-out " (see p. 101), which is then re-colonised in much the same way as the fresh sand on the newest dunes (Pl. XXI, p. 110). As another example, the seaward edge of a salt-marsh sometimes becomes eroded as a result of a sudden change of current or for some other reason. The original vegetation is thus destroyed, but in course of time the mud on which it originally grew may be colonised once more to form what is called " secondary marsh," usually at a different level from the original one.

When we come to look into the reasons why particular plants grow where they do, we find that there are a large number of factors to take into account. Most of these are closely inter-related in the effects they produce, but it is worth while to discuss briefly some of those which are especially important in determining coastal vegetation.

The climate of the country is obviously of the greatest importance, for it controls such factors as the duration and intensity of the sunlight, the range of temperature, the rainfall, the humidity of the atmosphere

Robert Atkinson

Plate VII   Open community of sea-rocket, *Cakile maritima*, etc., on a sandy
shore of the Isle of Berneray, Sound of Harris.

A. Sampson

Plate VIII   Hawthorn tree distorted by the prevailing south-westerly winds.   Bude, Cornwall.

and the strength of the winds. Climatic factors show their effect most clearly when vegetation is studied on a broad geographical basis, but even in a relatively small area like that of the British Isles the effects of small differences in climate are quite noticeable. Thus the average rainfall and the humidity of the air is much greater on the west coast than on the east, and this probably accounts for many small differences in the distribution of plants along the two coasts. It is certainly responsible for the much richer moss flora found on the western sand-dunes compared with those on the east coast, and may partly account for the occurrence of certain typical " Atlantic " species along our western and south-western coasts. In the same way, the mean temperature in the North is distinctly lower than that in the South, and this is one of the factors responsible for the absence or rarity of a number of plants in Scotland and northern England, which are comparatively common in the South, and also for the fact that certain characteristic north European plants are only found in the North.

Wind is obviously a very important factor in all coastal habitats. Its most pronounced effect is that it increases the loss of water vapour from the leaves of plants by constantly bringing dry air into contact with them. As a result, the growth of many seaside plants is considerably retarded and they are often found in a very stunted form. To combat this, many coastal plants adopt a mat or rosette habit for much of the year. Exposed parts of the coast are generally destitute of trees, and such few trees as do occur near the coast are usually found tucked away in sheltered valleys, or combes as they are called in the West Country. Trees and hedges in coastal areas often assume very distorted forms, which show clearly the direction of the prevailing winds (Pl. VIII, p. 31). This is due to the fact that only the shoots on the leeward side can develop normally, those continually exposed to the prevailing winds being dried off and killed. In this way they appear to have been blown over by the strength of the wind, whereas actually their peculiar shapes are due to the unequal development of the shoots on their two sides. The effect of wind in retarding growth is most marked on the east and north-east coasts, which are exposed to the driest winds, although it is very noticeable on any of our coasts.

Another group of factors to be considered depend upon the general topography of the habitat and may be called *physiographical factors*. The angle at which the ground slopes, the aspect or direction of the slope and the height of the land above sea-level, are examples of these.

The familiar coastal processes of erosion, silting and the blowing of sand, which are discussed in Chapter 2, also come into this class. In addition, the prevalence of strong winds along the coast, whose effects have just been described, is clearly due to a combination of climate and topography. It is hardly necessary to give illustrations of the result on the vegetation produced by all these factors; the relation of the highly specialised community of plants which are found on mobile sand with their habitat, for instance, is sufficiently obvious. Some of them, however, become particularly important when we consider cliff-vegetation. Thus the angle at which the cliffs slope largely controls the amount of soil available for supporting plants in the rock crevices, and will indeed determine the stability of the surface of the cliff itself, if it is composed of soft material. The height above the sea will also determine the amount of spray to which the habitat is exposed, and most cliffs show some zoning of the vegetation which can be correlated with this factor. The direction towards which a cliff faces is also important in determining the amount it will be exposed to the prevailing winds and thus, indirectly, the amount of spray it is likely to receive, and will also control the duration of the periods of shade. There is often a marked difference in the vegetation of cliffs with different aspects, in particular those on the opposite sides of small islands.

Another group of factors, which in some ways show the most pronounced effects on the composition of the vegetation, are those related to the physical and chemical properties of the soil. These are called *edaphic factors* (Greek: *edaphos* =the ground). On the coast the commonest physical characteristic of most habitats is that of a poor water-supply. Sand-dunes, shingle beaches, and most cliffs are all subject to periodical drought conditions, which are aggravated by the drying winds. We shall see in the next chapter that the leaves of many seaside plants are equipped with devices of various kinds to check excessive loss of water, and that their root-systems are often very extensive. The amount of air contained in the soil is also related to its physical state, and it is noticeable that a number of plants, such as marram-grass on dunes, sea purslane in salt-marshes, and the shrubby seablite (*Suaeda fruticosa*) on shingle grow luxuriantly only when their roots are well aerated.

The chemical nature of the soil is also of great importance. Salt is obviously the master-factor in determining the highly specialised vegetation of salt-marshes, and the presence of halophytes in other coastal habitats, such as shingle beaches and exposed cliffs, shows that there also salt spray is deposited in sufficient amount to be an important factor. The ultimate vegetation developed on sand-dunes also varies greatly with the amount of calcium carbonate initially present in the sand. We have already seen that, if this is small, it will be washed out of the surface-layers in time, and that typical plants of acidic soils like heather and heath may eventually appear, as the supply of humus increases. Many west coast dunes, however, have been formed from sand which contains so much calcium carbonate in the form of broken shells that the relatively slow leaching action of the rain has produced little effect on it. As a result, the final vegetation on these dunes remains fundamentally calcicole (lime-loving), and is remarkably similar in composition to the grassland commonly found on chalk and limestone. In the same way, chalk and limestone cliffs may be expected to show some different plants from those which are found on acidic rocks.

Finally, we must say a word about the effects on the habitat caused by living organisms. These are called *biotic factors* (Greek: *bios* = life), and include the activities of man and his animals, the effects of rabbits, birds and insects and those produced by the plants themselves. The effects of previous generations of plants in altering the physical and chemical properties of the soil have already been briefly discussed. As far as man is concerned, his activities are less in evidence along the coast than in most parts of the country, since coastal areas do not lend themselves well to agricultural development. Nevertheless, in a thickly populated area like ours, there is no region where the hand of man has not played some part in modifying the vegetation. For instance, large areas of many salt-marshes are used for the grazing of cattle, which has the effect of restricting some plants but not others. Many old salt-marsh areas, too, have been completely transformed by drainage operations or the construction of sea-walls to exclude the tides, and the laying out of golf-courses has altered the vegetation in sand-dune areas in a number of places. Moreover, in certain districts marram-grass has actually been planted to stabilise shifting sand-dunes, and elsewhere rice-grass has been employed in a similar way for reclaiming salt-marshes, so that it is often impossible to distinguish between natural and partly artificial vegetation. Nor should it be forgotten that the large-scale felling of the native forests all over the country in the past has had the indirect effect of preventing the natural development of the climax vegetation in many suitably undisturbed areas along the coast.

Rabbits are frequently responsible for considerable modification of the vegetation, and are often extremely common in coastal areas. In particular, the grassland on the tops of cliffs is often infested with them, and the older sand-dunes provide a veritable rabbit's paradise. In all probability, the somewhat stunted vegetation which is so characteristic of such areas results as much from its being continually nibbled by rabbits as from its exposure to strong winds. Some plants, however, are more attractive to rabbits than others, so the actual composition of the vegetation may be considerably altered. Even salt-marshes are not exempt from the attentions of rabbits; in some districts, for instance, it is unusual to see more than a quite small proportion of the sea-aster plants reaching the flowering stage. Birds also sometimes have a marked effect on the vegetation, particularly when large colonies gather on small islands for breeding purposes. Needless to say much excreta is deposited on the cliff-ledges and cliff-tops near their nesting sites, and the increase in the amount of nitrogen and phosphates in the soil produced in this way has the effect of altering the composition of the vegetation considerably.

The above brief summary can do no more than suggest the kind of factors which must be looked for if we are to make any attempt to understand why coastal vegetation is distributed as it is, and why particular species occur in some places and not in others. Our knowledge of these matters is still extremely incomplete, and it is well to realise that much valuable information can still be easily collected by amateur botanists who are prepared to make a fairly detailed survey of the vegetation in a particular habitat and to keep their eyes open for the factors which have been responsible for its composition.

John Markham

Plate 1   Sea-holly, *Eryngium maritimum;* a characteristic dune plant

CHAPTER 4

FORM AND
HABIT OF COASTAL PLANTS

THE MAJORITY of the plants we find growing round the coast have to contend with unusually harsh conditions, and many of them are specially adapted to enable them to survive in their inhospitable habitats. In this chapter we shall consider some of the characteristic growth-forms they adopt.

Undoubtedly the main problem for most of these plants is to obtain adequate supplies of water, particularly in the early stages of their growth. This applies both to those growing in such obviously dry habitats as sand-dunes, shingle beaches or rocky cliffs, and to those growing in saline ground, such as salt-marshes or brackish swamps, although the reason for the difficulty is quite different in the two cases. The whole question of water-supply is sufficiently fundamental to merit discussion in some detail.

To deal first with the dry habitats; the whole trouble here is that they do not retain sufficient quantities of water in their surface layers, since the " soil " they provide is largely made up of coarse particles. The water-holding power of a soil depends in the first instance on the size of its particles. If these are large, water can percolate easily through them, and will also evaporate more quickly because of the large air-spaces between them. Thus the greater the number of small particles, the longer the soil will take to become dry after rain. Furthermore, it is a well-known fact that water tends to stick on to the outside of all relatively small particles on account of the force known as " surface-tension," and since the total surface-area of a given weight of small particles is clearly greater than that of the same weight of coarse particles, the finer the soil the greater its powers of retaining water. But in addition to a lack of small particles, there is usually a shortage of humus in all the habitats in question. This important material, consisting of dead plant-remains in the process of decay, has already been briefly referred to (p. 28). Without discussing the varied forms in which this organic matter can occur, the amount present in a soil can be roughly guessed from its colour. Thus dark-coloured " peaty " soils contain the greatest amount and sands the least. All farmers are familiar with the fact that adequate quantities of this material are necessary in all " light " (i.e. coarse) soils, if they are not to suffer from frequent drought conditions. Humus possesses great powers of absorbing water, chiefly because much of it is usually in the form of very small particles of what are called " colloidal " size (i.e. they are so small that they easily pass through a filter-paper, and take a long time to settle when they are suspended in water). Quite apart from this, it is a valuable source of plant food, partly on account of the nitrogen it contains, but principally because it absorbs valuable salts and prevents them from being washed away.

The plants growing in saline habitats also have trouble with their water-supply, though of a very different kind. Here there is often an abundance of water, but it is, of course, salt water. As a result, the plants may suffer from what has been called a " physiological drought." This means that, despite an abundance of water in the soil, they are unable to make use of it on account of the high concentration of salt it contains. Many salt-marsh plants, therefore, live under conditions of partial drought rather similar to those encountered in other coastal habitats. As evidence of this, it can often be noticed that they are greatly benefited by the dilution of their soil-water, when a spell of wet weather occurs in the summer. Indeed many, though not all, halophytes can grow luxuriantly in ordinary garden soil.

Another characteristic of coastal habitats is that they are all to a greater or lesser extent exposed to strong winds. The most important result of this, as has already been pointed out, is to increase the rate of evaporation of water at the leaves (transpiration). This causes plants to draw further on their slender water-supplies, and if these are inadequate, wilting may take place. Thus wind aggravates the results of the water-shortage.

In order to understand the various ways in which maritime plants deal with this fundamental problem of water-supply, it is necessary to say a word about two processes, common to all plants, which are specially important in this connection. These are osmosis and transpiration.

Osmosis

All plants obtain the water and soluble salts required for their growth through the agency of cells situated near the ends of their roots, which are known as the root-hairs. These cells are filled with sap, which contains small quantities of soluble salts and much larger amounts of soluble organic substances, such as sugars, in solution. When the root-hairs are in close contact with the soil-water, a suction pressure is developed through the walls of these cells, called the osmotic pressure. As a result of this, water passes into the cell and temporarily dilutes the sap. This is a familiar chemical phenomenon and can easily be demonstrated in a number of ways. If any solution is enclosed in what is called a "semiper-meable membrane" (i.e. one which will allow water, but not dis-solved substances, to pass freely through it), the osmotic pressure of the solution will cause water to be sucked into the solution through the membrane.

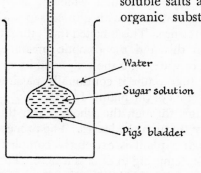

Water

Sugar solution

Pig's bladder

Fig. 2.—Experiment to demonstrate osmosis.

A simple way in which this can be demonstrated is to tie a piece of pig's bladder very firmly over the end of a small funnel (preferably a "thistle funnel"). If the funnel is now filled with a strong solution of (say) cane sugar and is immersed in a vessel containing water, the water will soon start to pass through the membrane into the sugar solution and rise up the stem of the funnel (Fig. 2). If a long piece of glass tubing is attached to the stem, it will be noticed, after a day or two, that a column of water several feet high has risen up the tube. It is important to realise that the height of this column is not a measure of the pressure exerted by the initial sugar solution, since this solution is naturally becoming steadily more dilute as the water enters it. In point of fact the osmotic pressure of a 15 per cent solution of sugar is about 10 atmospheres, or sufficient to support a column of water of well over 300 feet in height!

The walls of the root-hair cells function in exactly the same way as semipermeable membranes, though they allow considerable amounts of the substances dissolved in the root-water to pass through them also. Measurements of the osmotic pressures exerted by the cell-sap of many different plants have been carried out. These have been found to vary considerably, but a value of about 10 atmospheres for normal plants (mesophytes) can be taken as an average figure. As a result of this large suction pressure, it is obvious that a root-hair which is freely supplied with water will soon become distended with diluted sap and develop a corresponding balancing pressure. This is called the " turgor pressure." An equilibrium between this and the osmotic pressure, resulting in the cessation of the flow of water into the cell, would soon be reached were it not for the fact that water is continually passing from the root-hairs into the root and stem of the plant. It is this movement which maintains a flow of water through the whole plant, the excess water being eliminated largely through the leaves. The process by which water is conducted through a plant is extremely complex, and there is nothing to be gained by attempting to discuss it here. The important thing to understand is that, provided water is in contact with the root-hairs, a steady flow into the root can be maintained. On the other hand, if insufficient water is available in the soil, the plant may not be able to obtain an adequate supply to build up its turgor pressure, with the result that the whole plant becomes limp and is said to " wilt." Obviously this danger is greater if water is eliminated too rapidly by the leaves, and plants growing in dry habitats are often provided with devices to prevent excessive transpiration.

Measurements of the osmotic pressures exerted by both halophytes and xerophytes have been shown to be, in general, much larger than those characteristic of normal plants. We have mentioned 10 atmospheres as a typical value for mesophytes, whereas 40 atmospheres would appear to be an average value for plants in the former classes. Indeed, some desert plants have been shown to exert pressures running up to the enormous figure of 100 atmospheres and more. Obviously this greatly increased power of suction must be of much assistance to plants growing in arid soils in enabling them to obtain what little water there is. In the case of halophytes, a high osmotic pressure is

virtually essential if they are to overcome the considerable pressure
of the salt water in which they have to grow. Ordinary sea-water has
an osmotic pressure of about 20 atmospheres, but in a salt-marsh the
concentration of salt may become very much higher during a spell
of dry weather in those areas which are not submerged by every tide.
If halophytes were incapable of exerting a greater osmotic pressure
than that of the salt water in a marsh, osmosis would take place in the
wrong direction and water would be sucked out of the plant into the
soil-water. Thus the plant would not only fail to obtain its water-
supply, but would lose much of the water it already contained. This
effect can easily be demonstrated by putting any normal plant into
salt water, when it will be seen to wilt in a very short time.

A good deal of work has been done on the measurement of the
osmotic pressures developed by halophytes when growing in salt
solutions of varying concentrations. The results show clearly that they
fluctuate considerably and are able to alter rapidly to adjust themselves
to changes in the concentration of salt in the soil-water. This adapt-
ability accounts for the wide tolerance shown by many halophytes
growing in different parts of salt-marshes. The whole problem of the
mechanism by which water is absorbed by plants is very complex, and
the above account is much simplified in order to explain the main
principles.

TRANSPIRATION

Passing now to the other end of the plant, we must say something
about the process by which the surplus water is disposed of at the
leaves. This is known as transpiration. On the surface of any leaf a
number of minute pore-like openings are to be found which are called the
"stomata." Each stoma usually takes the form of a slit between two
elongated cells known as " guard cells," lying side by side (Fig. 3(c).)
The opening or closing of the pore is controlled by the swelling or
contraction of this pair of cells. Thus, when the turgor pressure of
the plant is high, the cell-walls expand and the slit is opened to aid
the elimination of water. When the water-supply is less abundant,
the turgor pressure falls and the cells contract so that they lie with
their walls in contact with each other, thus closing the slit. It should
be emphasised that the stomata are not only concerned with the
elimination of water-vapour but are also the organs through which

the plant absorbs carbon dioxide and gives out oxygen in the carbon assimilation process (photosynthesis). They are in fact the openings through which the exchange of all gases takes place, although to some extent the whole surface of the leaf and even the stem functions in this capacity. When the external covering or "cuticle" of the leaf is thick, however, the process is largely confined to the stomata. Usually these occur more thickly on the under-surface of the leaf, as being better protected from the drying influence of the sun. Only in water-plants with floating leaves are they confined to the top surface. Although the number of stomata found on the leaves of different plants varies greatly, there do not appear to be any fewer on those belonging to halophytes or xerophytes than on the leaves of normal plants. A moderately large leaf with an average density of stomata may possess several millions of such openings.

Although there is much which is obscure about the transpiration process, it has two important effects. Firstly, it maintains a constant flow of water from root to leaf through the wood of the plant, bringing with it also small quantities of dissolved salts which are essential for the plant's growth. Secondly, it tends to reduce the temperature of the leaf when it is exposed to the heat of the sun. It is a well-known fact that when a liquid is changed into vapour, energy (latent heat) has to be expended. This heat is derived from the air immediately in contact with the surface of the leaf and in this way the leaf itself is cooled. In hot climates and in dry habitats this result may be important. The chief danger with xerophytes, and to a lesser extent with halophytes, is that the loss of water by transpiration may be so rapid that it cannot be replaced from the scanty supply of water available at their roots. Many plants belonging to both these classes are therefore equipped with devices to check excessive transpiration, and some of these will now be described.

TRANSPIRATION-CHECKS
OR DEVICES FOR REDUCING TRANSPIRATION

The development of a thick cuticle or outer skin on the leaves is the simplest and most frequently adopted method for the reduction of transpiration. The leathery feel of the leaves produced by many seaside plants is a characteristic which can hardly be overlooked, though the development of thick cuticles is by no means confined to coastal plants. In some cases this thickening is supplemented by the secretion of wax on the leaf surface, as in the case of the sea-holly (*Eryngium maritimum*) (Pl. 1, p. 35). These protective layers have the effect of confining the evaporation of water entirely to the stomata, for in their absence a considerable amount of water is lost through the rest of the surface. Fig. 3 shows diagrammatically a transverse section round a stoma of a leaf with a thin cuticle (*a*) and a similar section from a leaf with a thick cuticle (*b*). It will often be noticed that the thickness of the cuticle varies considerably amongst individuals of the same species, according to the habitat in which they are growing. The leaves of the scarlet pimpernel (*Anagallis arvensis*) (Pl. 2b, p. 50), for instance, become thick and leathery when it is growing on bare sand amongst dunes, although under normal conditions in garden soil they are soft and slender.

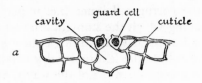
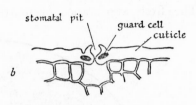
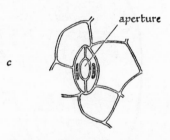

FIG. 3.—Types of stomata: *a.* Transverse section of leaf with a thin cuticle: *b.* Transverse section of leaf with a thick cuticle, showing a sunken stoma; *c.* Surface view of a stoma.

In many plants the stomata are protected by being placed in grooves or hollows sunk well below the surface of the leaf (Fig. 3 (*b*)). In the dune grasses, marram-grass (*Ammophila arenaria*) and sea lyme-grass (*Elymus arenarius*), the stomata are mostly confined to the bottom and sides of the deep grooves in their leaves. This protection is much improved by the tendency of the leaves to roll up into a narrow tube in dry weather, which has the effect of maintaining a layer of air, largely saturated with water-vapour, between the stomata and the outside air, and thus reducing evaporation. The manner in which air is enclosed when the leaf rolls up is clearly shown in Fig. 4, and the corrugated inner (i.e. upper) surface is due to the deep grooves along which the stomata are scattered. The outer (i.e. under) surface is furnished with a thick cuticle and is devoid of stomata. This habit of rolling the leaf under dry conditions is shared by many plants, and is a good example of the way they can adjust themselves to variations in their water-supply. When water is plentiful, the blade opens out and becomes flat, thus exposing a greater surface for transpiration. The fresh appearance of marram-grass on open sand-dunes after

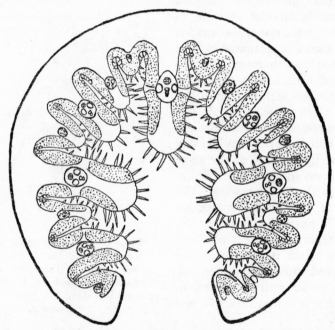

FIG. 4.—Transverse section of marram-grass leaf when rolled (from Fritsch & Salisbury, 1946).

abundant rain is quite distinct from its parched look after a long period of dry weather, and on closer inspection will be found to be due to the unfolding of its leaves.

Another common way in which the stomata are protected is by the growth of hairs on the surface of the leaf. These are often associated with sunken stomata and are very effective in maintaining a damp atmosphere round the opening, since moisture tends to condense on them. The stiff hairs protecting the furrows on the upper surface of the marram-grass leaf will be noticed in Fig. 4. Many seaside plants have hairy leaves, and some are covered with a thick down. The yellow horned poppy (*Glaucium flavum*) (Pl. IX, p. 54), the sea stock (*Matthiola sinuata*) and the buck's horn plantain (*Plantago coronopus*) (Pl. XXXVI, p. 179) are good examples of coastal plants with hairy leaves, while the leaves of sea-wormwood (*Artemisia maritima*) (Pl. XXXI, p. 166) and the tree-mallow (*Lavatera arborea*) are markedly downy. The characteristic silvery foliage of the sea-buckthorn (*Hippophae rhamnoides*) (Pl. XX, p. 103) and sea-purslane (*Halimione* (*Obione*) *portulacoides*) (Pl. 16, p. 210) is also due to scale-like (peltate) hairs covering the surface of the leaves. These hairs

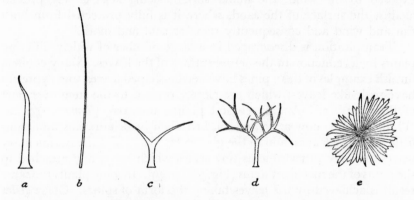

a   b   c   d   e

FIG. 5.—Typical covering-hairs on various leaves: *a. Plantago coronopus; b. Cynodon dactylon; c. Erophila verna; d. Matthiola sinuata; e. Hippophae rhamnoides.*

are usually dead when the leaf is mature, and contain only air. Apart from aiding the retention of moist air near the surface, they reflect much of the sun's heat. Some leaves possess simple un-branched hairs, but those on many others are branched and occur in very different forms. Some typical covering hairs from the leaves of coastal plants are shown in Fig. 5. Like the thickness of the cuticle, the degree of hairiness shown by individuals of the same species often varies with availability of the water-supply in the habitat. Thus the sand-dune form of silverweed (*Potentilla anserina*) commonly shows a thick felting of silvery hairs on the upper surface of its leaves, as well as on the lower.

Still another way in which relatively damp air is maintained over the surface of the leaves is by the plant adopting a dense mat habit, so that the transpiring surfaces of the leaves are kept in close contact with each other. Alpine plants often mass their foliage in this way, but amongst coastal plants thrift (*Armeria maritima*) provides one of the best examples, since its habit varies considerably with the place in which it is growing. Thus the close rosette form is typical when it is growing on rocky cliffs and other dry habitats, or when it is heavily grazed, whilst with a better water-supply it assumes a much more open habit (Fig. 6). Many sand-dune plants spend most of the year in the form of a rosette, only sending up a vertical stem during the flowering season. In this way, only the upper surface of the leaf is exposed to the wind, the under surface being kept closely pressed against the surface of the sand, where it is fully protected from both sun and wind and consequently remains cool and moist.

Transpiration is discouraged in a large number of widely differing plants by a reduction in the actual surface of the leaves. Many conifers furnish examples of this; pines have needle-shaped leaves, and cypresses have scale-like leaves, which are closely pressed to the stem over part of their surface. Among coastal plants, tamarisk (*Tamarix gallica*) (Pl. 8, p. 126), now a well-established alien in Britain, has numerous little scale-like leaves, and in the glassworts (*Salicornia*) the rudimentary leaves are only just visible as tiny scales which are firmly attached to the joints of the succulent stems (Fig. 9(*b*) p. 49). In some plants the same result is achieved by the leaves taking the form of spines. Gorse (*Ulex* spp.) is the best-known example in this country, but in desert regions the majority of the xerophytic plants show this modification, the *Cacti* being a familiar case. Our native xerophytes more frequently develop spiny margins to their leaves, thistles furnishing the obvious example. The most striking seaside plant to show this development is the sea-holly (Pl. 1, p. 35), though the leaves of the prickly saltwort (*Salsola kali*) (Pl. I, p. 6) also terminate in stout spines. The tendency to form woody tissue in the form of spines appears to be closely related to a shortage in the water-supply. A number of plants which produce spines when growing in dry habitats do not possess any when moisture is abundant.

Occasionally the function of the leaf is taken over by specially modified branches known as "cladodes." The only coastal plant exhibiting this modification is the wild asparagus (*Asparagus prostratus*),

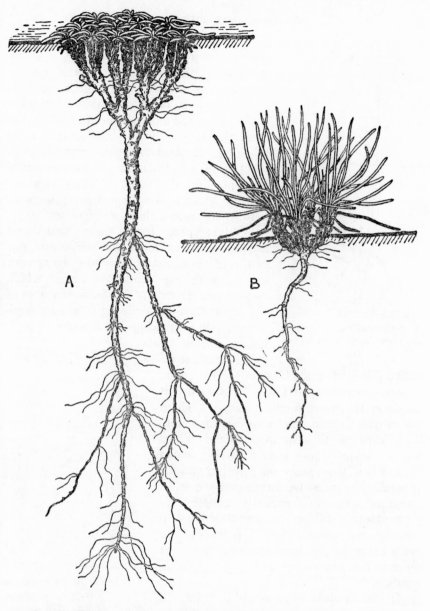

A   B

Fig. 6.—Different forms adopted by thrift: *a*. Rosette form under grazing or in dry ground; *b*. More diffuse habit when protected from grazing and with a good water-supply (from Tansley after Yapp, 1917).

FIG. 7.—Part of a branch of asparagus, showing cladodes and scale-leaves (*s*).

a rare plant found on sandy shores in a few localities only in this country. If the familiar feathery foliage of the garden asparagus is examined, it will be seen to consist of tufts of short leaf-like branches arising from the axils of minute scaly leaves (Fig. 7). It is difficult to see exactly what advantage a plant can gain from the substitution of a leaf-like stem for an ordinary leaf—possibly the tissue of the cladode is more resistant to shrinkage when the plant is suffering from a shortage of water.

Quite apart from these permanent alterations in leaf-form, the shape and size of the leaves of many common plants vary greatly with the conditions under which they grow. For example, the first leaves of the red-fruited dandelion (*Taraxacum laevigatum*), when growing in a moist hollow among sand-dunes, are often quite entire (i.e. with smooth edges); later in the season, when the sand has become dry, it produces the more familiar deeply divided leaves with a much smaller surface-area (Fig. 8). Most of the common inland plants found on sand-dunes possess smaller leaves than when they grow on more hospitable ground. Nor must we forget that the semi-prostrate form so frequently adopted by dune-plants is still another method by which excessive transpiration can be reduced, since every extra inch in height exposes the plant more to the desiccating action of the strong winds.

It will be clear from what has been said that many plants are capable of modifying their normal form when growing in dry habitats. Of the various transpiration-checks

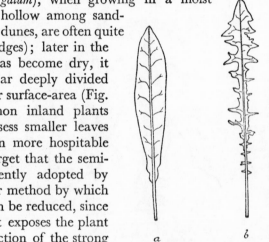

a   *b*

FIG. 8.—Different leaf-forms of the red-fruited dandelion: *a.* Young leaf from a plant growing with abundant moisture; *b.* Leaf from a plant growing on dry sand.

which have been described, undoubtedly the development of a thick cuticle is the most frequent one employed by coastal plants. It possesses an added importance for plants inhabiting open sand-dunes in that it also protects them from possible injury caused by the sand being blown against them. Anyone who has done any botanising on exposed sand-dunes during a high wind will know how violent this bombardment can be!

It is important to point out, before we leave this subject, that it is only during periods of water-shortage that these mechanisms for reducing transpiration become important. Recent research has shown that xerophytes transpire during wet spells at least as much as, and often more than, ordinary plants. In those cases where the transpiration-rate becomes unusually high, it may be related to the necessity for rapid growth and carbon assimilation during the infrequent wet periods. What really characterises a xerophyte is that it *can*, if need be, decrease its transpiration-rate to a minimum when living under drought conditions. In addition, the actual protoplasm (living matter) seems able to withstand desiccation to an unusual extent.

SUCCULENCE

Some xerophytes employ quite a different method to provide against water shortage, though it is often found in combination with the leaf-modifications already described. It will be noticed that many plants growing in dry places have a fleshy or succulent appearance. This is due to the development of large colourless cells, known collectively as " aqueous tissue," which are employed for storing water. This is usually confined to the leaves as in the stonecrops (*Sedum* spp.) (Pl. 14, p. 190) or the sea-spurge (*Euphorbia paralias*) (Pl. XIX, p. 102), but sometimes the whole stem is succulent as in the glassworts (*Salicornia* spp.) (Fig. 9(*b*)) or the familiar *Cacti*, the leaves in these cases being reduced to mere scales or spines. As a rule these cells occupy the centre of the leaf or stem, and the green cells which are used for photosynthesis occur nearer the edges. In dry weather, as water is gradually lost by transpiration or by its passage into the green cells, the water-holding cells shrink; when the water-supply improves, they expand once more. They function, in fact, as water-storage cisterns for use by the plant in times of drought. Desert succulents, which often possess extremely thick cuticles to reduce transpiration, can exist for prolonged periods without an external supply of water, during which they gradually shrivel until they can replace their internal water supplies when the rain comes.

It is rather surprising that halophytes should form the largest class of plants exhibiting succulence in this country. It has already been pointed out that the chief characteristics of this group are that they can exert sufficiently large osmotic pressures to withdraw water from a soil which is saturated with sea-water, and that the protoplasm forming their cells is not injured by exposure to salt solutions. Under normal conditions, therefore, they should not encounter much trouble with their water-supplies, and it is difficult to understand why they should develop aqueous tissue so extensively. It is possible that they draw on their internal reserves of water when the concentration of salt in the soil alters too rapidly for them to accommodate their osmotic pressure to it, but this can hardly account for such a wide-spread characteristic. Furthermore, it has been shown that halophytes do not, in fact, withstand drought like succulent xerophytes, but actually wither quite rapidly.

The most likely explanation is that the similarity in appearance of succulent xerophytes and halophytes is largely accidental. There is considerable evidence to suggest that the latter become succulent as a result of some chemical effect associated with salt, probably with the chlorine rather than the sodium part of it (salt is a simple compound between these two elements). All plants when growing in a saline soil absorb some salt, for the cell-walls of the root-hairs never function as " perfect " semi-permeable membranes, but allow a certain amount of the dissolved substances to pass through them. This is, of course, true of all types of plant, for otherwise they would be unable to obtain the small quantities of other mineral salts essential to their growth. Many halophytes, however, absorb very large amounts of common salt; the ash of some of them, like the glassworts, was formerly used on a large scale to provide soda for glass-making, and certain plants, as for example thrift, actually excrete surplus salt from the glands on their leaves. That the development of succulent leaves is closely connected with the absorption of salt is borne out by the behaviour of many non-halophytes when they grow in places exposed to sea-water. Many inland plants found on open beaches or on cliffs within reach of sea-spray possess much more fleshy leaves than they have in their normal habitats; bird's-foot trefoil (*Lotus corniculatus*), kidney vetch

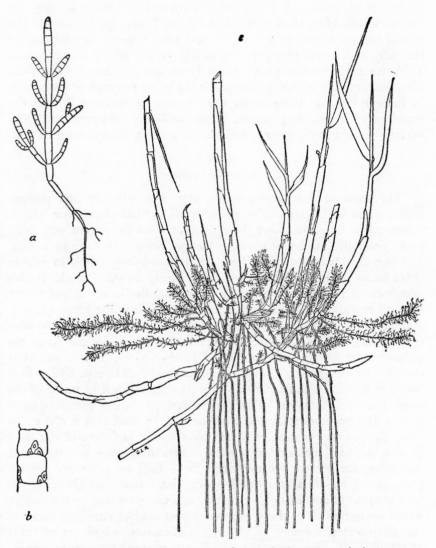

FIG. 9.—Comparison of root-systems of annual glasswort and rice-grass:- *a*. Small plant of annual glasswort; *b*. Two joints from stem of annual glasss wort enlarged to show leaf-scales; *c*. Base of a rice-grass plant showing vertical anchoring roots and horizontal feeding roots and stolons (Fig. *c* from Tansley after Oliver, 1926).

(*Anthyllis vulneraria*) and the greater knapweed (*Centaurea scabiosa*) are species which often show this effect. Some years ago I analysed the ash of certain inland plants which had been exposed to sea-spray in this way and found 13.5 per cent of salt in that of the kidney vetch. Evidence such as this points strongly to the conclusion that some chemical action connected with salt is the primary cause of succulence in halophytes, and there seems little reason to associate it with the problem of conserving water. The similarity between succulent xerophytes and halophytes is remarkable and we must leave it at that.

ROOT-SYSTEMS

The roots of coastal plants are very characteristic and present many points of interest. The majority of *true* halophytes (i.e. plants which normally grow where the soil-water is saline) possess very deep roots, generally markedly woody. Most of these plants are perennials and the chief value of their long roots in a salt-marsh is to enable them to secure a firm anchorage in relatively unstable mud. It also allows them to derive their main water supplies from regions where the concentration of salt is less variable than it is in the surface layers. Annual glasswort (*Salicornia stricta*) (Pl. XIII, p. 66) possesses only quite short roots, and as a result is liable to become dislodged from the unstable mud if there is a strong tidal flow (Fig. 9(*a*)). This is in marked contrast to rice-grass (*Spartina townsendii*) (Pl. XIV, p. 67), which occupies much the same position as a pioneer colonist in many of the south coast salt-marshes. This plant develops a most extensive root-system and becomes so firmly anchored in the mud that it can easily resist the strong currents produced by the ebb and flow of the tides (Fig. 9(*c*)). Its remarkable powers of spreading over soft mud and stabilising the surface are described more fully on page 71. Plants growing on sea-cliffs also develop very long roots, which serve the dual purpose of anchoring them firmly against uprooting by the violent winds encountered in these exposed places and of enabling them to tap deep-seated supplies of water. Well-established plants of thrift or samphire (*Crithmum maritimum*) (Pl. 10, p. 142) frequently possess roots several feet long, which penetrate deeply into the crevices between the rocks.

Extensive root-systems are also a characteristic feature of xerophytes in all parts of the world. In sand-dunes and shingle this enables the

a.   *John Markham*

b.   *John Markham*

Plate 2   *a.* Prostrate form of bittersweet, *Solanum dulcamara* var:
              marinum, growing on shingle. Essex
           b. Dune form of scarlet pimpernel, *Anagallis arvensis,*
              with fleshy leaves, Blakeney. Norfolk

plants to utilise the moisture which is always present some way below the surface (see p. 102). In addition, the elaborate root-systems developed by many dune plants perform the important function of binding blown sand. All pioneer colonists on sand-dunes, and to a lesser extent those on mobile shingle, have to contend with the possibility of being periodically swamped by loose sand or shingle. Most of them have, in varying degrees, the ability to form fresh shoots easily when they are submerged in this way, and to grow up through this covering. Marram-grass easily outstrips all other plants in the vigour with which it can do this. When once established in loose sand, it soon produces a mass of underground runners from which new shoots continually spring. Where these new shoots occur, fresh adventitious roots are produced under them; as the stems and leaves become buried by sand, further shoots and leaves are produced at a higher level on the stem (Fig. 10). The thick tufts of leaves and young shoots are very stiff and offer a considerable obstruction to the wind, causing it to drop some of its load of sand round them. Provided the blown sand does not accumulate too rapidly over the plants, marram-grass can continue to grow upwards through many feet of sand.

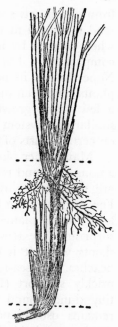

Fig. 10.—Leaf production at different levels shown by marram-grass (from Fritsch & Salisbury, 1946).

In this way dunes rising to considerable heights can be produced, their size depending on the supply of available sand and the strength of the prevailing winds. The whole interior of the dune remains closely penetrated by a mass of rhizomes (underground stems) and fine roots which bind it together. As these rhizomes and roots are constantly being renewed at higher levels, the lower ones gradually die off, but their dead remains persist in the lower regions of a dune for a long time and continue to exercise a stabilising effect on it (Pl. XXI, p. 110).

To a lesser extent the sea couch-grass (*Agropyron junceiforme*) can produce a similar result. This plant has much the same habit as ordinary couch-grass or twitch (*Agropyron repens*), which is an all too familiar agricultural weed. Like marram-grass, it produces a mass of rhizomes

from which new shoots spring up at frequent intervals, but its runners
tend to spread more rapidly in a horizontal direction than vertically
when covered by loose sand. As a result, the dunes it produces are
comparatively low compared with those formed by marram-grass
None the less, its powers of binding sand are considerable, and single
plants have been shown to cause dunes as much as 20 feet across during
a few years' growth. The sea lyme-grass (*Elymus arenarius*) has a
similar root-system and occasionally forms low dunes of the same type
on certain parts of the coast.

Many other pioneer plants on shifting sand can accumulate small
amounts of sand round them to form miniature dunes, provided they
are sufficiently virile to shoot up again when they become buried.
For example, the sea-sandwort (*Honckenya* (*Arenaria*) *peploides*) is a
low-growing plant not more than a few inches high, but it possesses
surprisingly extensive creeping roots and readily produces fresh
shoots when it is covered. Even those common pioneers of sandy
beaches, the sea-rocket (*Cakile maritima*) (Pl. VII, p. 30) and the
prickly saltwort (Pl. I, p. 6), which are only annuals, can collect
tiny dunes round their long trailing branched stems. Their dead
remains usually persist for a considerable time in the winter and
continue to hold the sand, although the principal agent here is the
stem rather than the roots.

Marram-grass has relatively little stabilising effect on the surface
sand, and it is the later colonists which establish themselves between
the clumps that are responsible for its eventual consolidation. Notable
amongst these are the sand-sedge (*Carex arenaria*) and the sand-fescue
(*Festuca rubra* var: *arenaria*), both of which produce horizontally
creeping roots just below the surface. Tufts of foliage arise from these
runners at frequent intervals, and in the case of the former often
appear spaced out along a nearly straight line for many feet, showing
clearly the course of its immense roots (Pl. X, p. 55).

Plants growing on shingle similarly have to endure periodical
swamping by stones. Very unstable shingle is usually devoid of
vegetation, but during storms those less mobile parts of the beach,
which carry quite a large amount of vegetation, may also be disturbed.
The shrubby seablite (*Suaeda fruticosa*) (Pl. XXIII, p. 114), a local plant
which is found on Chesil Bank and on some shingle beaches along the
Norfolk coast, adapts itself in an interesting way to these conditions.
When it grows in perfectly stable shingle it is a plant of erect habit,

3-4 feet high, but where the shingle is more mobile it assumes a quite different, semi-prostrate habit. Under the latter conditions, the stem is tilted forwards as shingle flows over it from above and the lower part becomes imbedded. New branching shoots then arise from the buried stem, and at the same time fresh tufts of roots are produced under each shoot. When growing on very unstable shingle banks, the original axis of many of these plants may become quite horizontal,

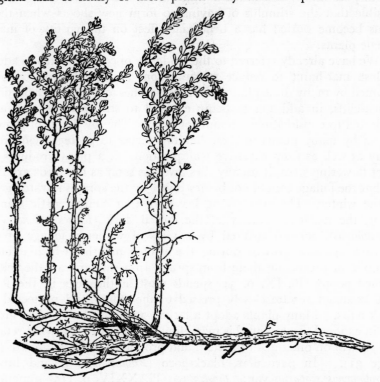

Fig. 11.—Typical habit of the shrubby seablite, when growing in shingle (from Tansley, after Oliver & Salisbury, 1913; drawing by Sarah M. Baker).

and a complex system of prostrate branches, covered everywhere with fibrous roots, will be found in the form of a dense mat just beneath the surface. The aerial shoots from these extensive buried stems grow vertically upwards, usually to a height of a foot or two, but much of the elaborate underground structure is generally dead and already in the process of decay (Fig. 11).

Several other characteristic shingle-plants are capable of enduring a certain amount of shingling-over. In particular, it has often been noticed that when the surface of the shingle has been unusually disturbed by storms during the winter both the sea-campion (*Silene maritima*) (Pl. XXXIII p. 174) and the sea-sandwort seem to grow more vigorously and flower more freely during the following season. The shrubby seablite is also much happier in mobile shingle, and it seems possible that the stimulus of having to form new shoots when their stems become buried has a beneficial effect on the virility of many shingle plants.

We have already referred to the tendency of coastal plants to adopt a close mat-habit to reduce transpiration, and to the rosette form assumed by many dune plants. Shingle beaches and the lower portions of sea-cliffs, in addition to being exposed to strong winds, are also subjected to considerable amounts of spray. The low habit of growth shown by many plants in these habitats serves to protect them from spray as well as from excessive transpiration. If a plant produces an erect flowering stem, it usually dies away as soon as the seeds are ripe so that the foliage should not be destroyed by the spray-saturated gales of the winter. Thus the young leaves of that characteristic shingle plant, the maritime variety of the curled dock (*Rumex crispus* var: *trigranulatus*), remain covered by a mass of withered foliage of the previous season's growth during the winter months, which is often effective in protecting them from spray. In a similar way, the yellow horned poppy (Pl. IX, p. 54) spends most of the year in the form of a compact rosette closely pressed to the stones and protected by thick hairs. Many plants adopt a largely prostrate habit when growing in exposed places, but this may often be caused by the wind retarding growth on one side of the stem and producing unequal growth (see page 31). In particular, blackthorn (*Prunus spinosa*) and broom (*Sarothamnus scoparius* var: *prostratus*) (Pl. XXIV, p. 115) sometimes grow completely flat along the ground on shingle, and a particularly well-marked prostrate variety of the woody nightshade (*Solanum dulcamara* var: *marinum*) (Pl. 2a, p. 50) may be seen on some south coast beaches.

Dune Annuals

To conclude this chapter something must be said concerning the highly characteristic collection of ephemeral plants which often appear

John Markham

Plate IX   Yellow horned poppy, *Glaucium flavum*, showing its long seed-pods and hairy leaves.

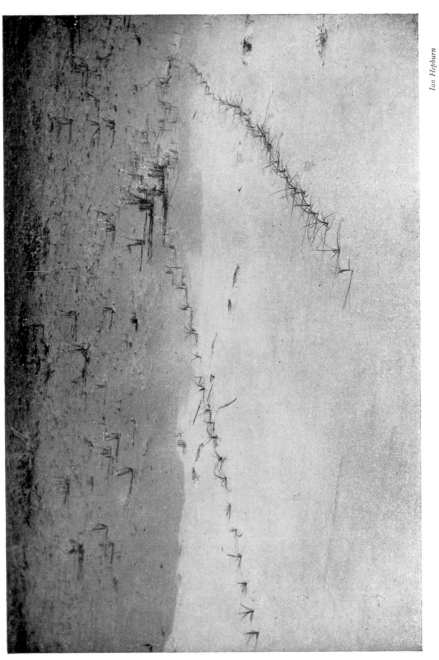

Ian Hepburn

Plate X Sand-sedge, *Carex arenaria*, colonizing a " blow-out " and showing the characteristic straight course of its horizontal roots. Kenfig Burrows, Glamorgan.

on the relatively bare sand of partially fixed sand-dunes. Very few of these species are confined to the coast and most of them are liable to turn up in waste ground anywhere. In marked contrast to the majority of perennial dune plants, these annuals have quite shallow roots, and a depth of more than 6 in. is unusual. A diagram of the root systems of some typical dune annuals is shown in Fig. 12 (N.B. 15 cms. = 6 in). These roots are devised to utilise the surface-water only, and it may be pointed out that such small amounts of humus as have accumulated at this early stage will be found near the surface.

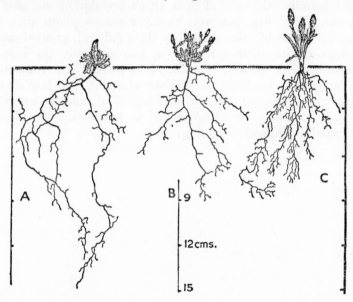

FIG. 12.—Root-systems of dune-annuals (from Fritsch & Salisbury, 1946). a. *Myosotis hispida* ; b. *Cerastium semidecandrum ; Phleum arenarium.*

In this connection it is worth mentioning that these " ephemerals " often occur more thickly near rabbit droppings, which have locally provided some additional humus. The surface-sand rarely remains moist sufficiently long during the summer months to enable such shallow-rooting plants to exist, and they aim at completing their life-cycle while it is still damp from the winter and spring rains. They may be described as " drought-escaping " plants, but are more usually referred to as " winter annuals," since they germinate in the autumn and persist as small seedlings through the winter. In early spring they start to grow quickly, then flower, and finally set their seed before the heat of the summer makes their habitat untenable. Their life-cycle comes to a rapid end with the setting of their seeds, and by the middle of the summer only their dried-up remains can usually be found.

Winter annuals are much more in evidence on the surface of dunes in some years than in others. Should an unusually long dry spell of weather occur in the late autumn, or should the surface be seriously disturbed by a series of especially violent gales when the seedlings have just germinated, most of next year's population may well be blotted out. They are, however, mostly common plants with light seeds, which are widely distributed by the wind and, granted favourable conditions for their germination, can generally be found in abundance in a bare habitat like this, where there is little competition from taller plants. Typical examples are whitlow-grass (*Erophila verna* agg.), early forget-me-not (*Myosotis hispida* (*collina*)) and the mouse-ear chick-weeds (*Cerastium semidecandrum* and *C. tetrandrum*). A fuller list will be found in Chapter 7 (p. 105).

CHAPTER 5

SALT-MARSH VEGETATION

I N THE following six chapters the main types of coastal vegetation
will be described in detail and this survey can best be opened by
considering the vegetation found in salt-marshes, since it is the most
completely maritime with which we shall have to deal. The plants
growing on shingle beaches and sandy foreshores may occasionally be
submerged for short periods by exceptional tides and many other
coastal habitats, such as sea-cliffs, are exposed to varying quantities
of salt spray, but salt-marshes alone are regularly inundated by sea-
water. It is this factor which differentiates them from all other habitats
and, as a result, practically the whole plant-population consists of
halophytes, for only these highly specialised plants can stand up to
regular exposure to salt water. Such non-halophytes as occur are
usually confined to the highest portions of the marshes, which are out
of range of all but very high spring tides. It is therefore not surprising
that the vegetation lacks the richness and variety of that found, for
instance, on sand-dunes, and in many places consists of a very small
number of different plants. Crabbe, who is apt to stress the grimmer·
side of nature, describes it thus:

> Here a grave flora scarcely deigns to bloom
> Nor wears a rosy bloom, nor sheds perfume:
> The few dull flowers that o'er the place are spread
> Partake the nature of their fenny bed.

Despite the lack of variety in the individual plants, I think many
naturalists find that salt-marshes have a peculiar charm of their own,
and will feel that Crabbe does them an injustice. For my own part,

ever since my earliest explorations of the Norfolk marshes near Cley and Blakeney as a schoolboy, I have always found them quite irresistible. It is rather difficult to analyse where their charm lies, for superficially they are not immediately attractive. They are, for instance, completely flat and conspicuously lacking in detail—as Crabbe says truly: "*Nor hedge nor tree conceals the glowing sun.*" Unless they occur at the mouths of estuaries with some hills round them, or there is at any rate some higher ground visible not far away, they can be the most desolate places in the wrong sort of weather. Moreover, they are generally rather irritating places to explore, since one can rarely walk in a straight line in any direction without having to cross an endless number of creeks, usually well supplied with thick mud. Nor is the strong smell of decay, which is always present, to everyone's taste, though to devotees it will immediately conjure up the charm of this type of country! Perhaps the very unfamiliarity of the vegetation is responsible for much of this charm, for it is unlike that in any other habitat. For the rest, the movement of the tides up and down the creeks and the slow seeping away of the water from the whole surface, which goes on ceaselessly, gives one the exciting feeling that the whole area is very much alive and in a state of active change. Nor must the added attraction frequently provided by the presence of large numbers of birds be overlooked, for surely few botanists will keep their eyes glued to the ground when there may be unusual waders and duck about. Even those who can hardly distinguish one bird from another will agree that the distant cries of gulls, oyster-catchers and curlews make a pleasant background to their more serious botanical studies and are part of the inherent attraction of this sort of country. But at certain times of the year salt-marshes can become very beautiful places in themselves and produce magnificent displays of colour. Some of the common plants cover large areas and grow in dense masses, and when, for instance, the sea-lavender, thrift, or even the sea-aster, comes into bloom, this rather dull-looking scenery suddenly bursts into its full glory.

Two words of warning are perhaps necessary to those who are not accustomed to wandering about on salt-marshes. Firstly, it is essential to be suitably dressed; either shorts and bare feet, or better, an efficient pair of waders is recommended, and even then some care in crossing the broader creeks is advisable if one is to avoid getting stuck or sitting down unexpectedly. Secondly, one should always ascertain the time of high-tide, for the water can run over the flat ground surprisingly fast, and the main creeks fill up even more quickly and soon become too deep to cross, so that there is a serious risk of being cut off.

There does not seem to be the same profusion of local names for salt-marshes as one finds used for sand-dune areas. The word " saltings " is widely used, but more often in connection with the salt-pasture developed on the higher levels and extensively used for grazing in many districts. Along the Severn estuary these upper marshes are often known as " Wharf-land." In some parts of Scotland and North England, however, the expression " Merse-lands " is used for the whole area.

There is one important characteristic which is shared by sand-dunes and salt-marshes alike, that they owe their very existence to the growth of plants. In both cases the production of a habitat which is suitable for the growth of the later colonists depends almost entirely on the initial efforts of the pioneer plants. On sandy or muddy sea-shores, which are fully exposed to wave-action, practically no vegetation, apart from seaweeds, can grow and flowering plants are usually confined to those portions which are out of reach of all but the highest tides. Salt-marshes, therefore, cannot develop unless there is some protection from the effect of the waves or the scour of strong tidal currents. They usually originate behind shingle or sand-bars, or in the upper sheltered regions of estuaries and bays, where headlands may provide sufficient protection. The areas of salt-marsh found in the British Isles vary greatly in extent, from the vast flat wastes round the mouths of the Thames or Humber, along the Solway Firth or bordering the Wash, to quite insignificant little strips along the estuaries of small streams or at the heads of sea-lochs on the west coast of Scotland. In their essential features, however, there is a considerable degree of similarity in the vegetation encountered in all examples of this formation and the fundamental processes involved in their production are the same everywhere.

How a Salt-Marsh is Produced

Before describing the vegetation in detail, it is necessary to give an outline of the usual stages by which the bare mud is transformed into the relatively stable vegetation found along the landward side of most marshes, and to explain how the plants which are found growing are related to the different levels of the ground. Fortunately most areas of salt-marsh can show some portion which is in active growth, and so one can often see the whole fascinating process literally going on before one's eyes.

All salt-marshes owe their origin to the fine mud, silt or sand which has been carried in suspension by the tide into some sufficiently sheltered area. In the earliest stages, one can sometimes watch a flat sheltered area becoming slowly submerged by the tide as it flows over it, and then almost imperceptibly reappearing as it ebbs. On a still day there may be scarcely a ripple on the surface, but the movement of the water will be just sufficient to keep most of the solids in suspension. At high-water, however, the tidal currents slow down to a minimum and there will be a short period when the water is relatively still (Pl. XIa, p. 62). It is during this quiet time that the deposition of solids principally takes place. Many salt-marshes have grown up on top of sand-flats and it is sometimes quite easy to see, by means of a little digging, that there is only a relatively thin layer of mud lying over a sub-stratum of sand. Sometimes a considerable amount of sand as well as silt is carried in by the tides, and these are then deposited together, particularly along the west coast. There are often seaweeds (marine algae) growing on the surface of sand or mud-flats and these play an important part in collecting silt, by offering some obstruction to the currents, and make some contribution to its subsequent stabilisation, though in times of storm they are liable to be uprooted. Besides carrying suspended solids, the tidal waters bring with them the seeds of various halophytic plants, and these are deposited at the same time as the silt. If the conditions are suitable for their germination, some of these may take root and, when they have developed, will be considerably more efficient than the algae in trapping silt. In this way, as more and more silt is deposited, the general level is very slowly raised, usually rather unevenly, and the tidal waters, instead of covering the whole area at once, begin to be divided by low islands and hummocks.

With the further deposition of solids, the islands grow in size, extending gradually and irregularly outwards from the inner edge of the marsh. The tides thus become more and more restricted to wide channels, and we can observe the earliest stage in the construction of the drainage channels and creeks which are such a prominent feature in mature marshes. The incoming water tends to take a winding course round the largest hummocks, and these twisting routes usually survive in the later phases (Pl. XIb, p. 62). As the flood water becomes more confined to these shallow channels, they are gradually deepened by the flow of the water up and down them. The tide often runs up the channels with considerable force, but the ebbing tides are probably more effective in scouring out the mud. Anyone who has wandered about on salt-marshes will be familiar with this rush of water up and down the creeks, and Crabbe describes it excellently when he says:

With ceaseless motion comes and goes the tide
Flowing, it fills the channel vast and wide.
Then back to sea, with strong majestic sweep
It rolls, in ebb yet terrible and deep.

As a result of the steady increase in height of much of the ground, the number of times the whole area is completely submerged during the month is reduced, and it gradually becomes possible for other halophytes, which are rather less tolerant towards immersion in salt water, to gain a footing on the higher levels. The order in which these plants appear varies considerably from place to place, and depends to some extent on the nature of the substratum (i.e. whether it is sandy, firm mud or deep, soft mud). With the arrival of these plants, the rate at which the solids are deposited by the tide is usually greatly increased, since they are larger and offer a much greater obstruction to the currents. The higher tides still overflow from the creeks and cover most of the surface of the marsh, so that the general rise in level is more rapid at this period in a marsh's history than at any other. Before long quite a close mat of vegetation will be established, whose average height may be as much as 6 inches above the surface of the mud. At this stage, the creeks begin to look more permanent, and their banks often become slightly raised above the general level owing to the vegetation being thicker along their edges.

In course of time, the number of general inundations by the tide continues to get less, and the rate of upward growth naturally falls off correspondingly. It continues slowly, however, since the whole area will be periodically submerged by spring tides, and a considerable amount of humus will also be added from the decayed remains of the now extensive vegetation. In these later stages, a number of other plants, which can tolerate only occasional submergence in sea-water, commonly appear, and at the same time some of the pioneer plants disappear or become less prominent, as the area becomes drier. In many salt-marshes a zone dominated by certain maritime rushes will be found at the highest level. This represents the most advanced type of vegetation which is normally produced, without the complete exclusion of sea-water from the area. Only exceptional tides can reach this zone and this is the only part of a salt-marsh where inland species can be found in any number. If considerable amounts of fresh water come in at the top of a marsh, a fresh-water swamp may sometimes be formed, and one can thus witness a continuous transition from maritime to inland vegetation.

Salt-marshes may be formed either by growth outwards from the shore-line, or by inward growth from the shelter of a shingle or sand bar. The general principles governing their growth are the same, though those developing towards the shore generally take longer to reach maturity. The reason for this is that the protective bars can easily be inundated by exceptionally high tides or during storms, with the result that the early colonists in the area behind them are uprooted and washed away, if the level has not been sufficiently raised. Needless to say, in suitable places both outward and inland growth sometimes takes place simultaneously.

SALT-PANS

A conspicuous feature of all salt-marsh areas is the large number of depressions, unconnected with the drainage channels and often largely devoid of vegetation, which are dotted all over the surface (Pl. XII, p. 63). These are called " salt-pans " and have normally originated in two different ways. In the first instance, they may be caused by the irregular spread of vegetation during the quite early stages of colonisation. The hummocks or islands, which are first formed, may sometimes coalesce in such a way as to enclose a bare surface and so hinder the draining away of the water when the tide recedes. If the drainage outlet eventually becomes completely obstructed, a damp hollow will be enclosed, which may persist for an indefinite period. At spring-tides it will fill up with water, and during the neap-tides it will slowly evaporate and may become a very strong salt solution. In the shallower depressions the water sometimes dries up completely and crystalline salt is visible on the surface of the mud.

a   *V. Robertson*

b   *V. Robertson*

Plate XIa   High-tide in a salt-marsh, showing islands of unsubmerged
vegetation. Horsey Island, Essex.

b   General view of the channels and salt-pans at low-tide.
Horsey Island, Essex.

Plate XII Salt-pans in a salt-marsh; Blakeney Point in distance. Morston, Norfolk.

It is therefore not surprising that they are largely bare of vegetation, for the great variations in the salt-content of the soil, coupled with the lack of drainage, inhibit the germination of most plants. Occasionally it may happen that a creek cuts back into a pan and in this way provides it once more with a drainage outlet. When this occurs, vegetation will rapidly spread into it and eventually it will disappear. Sea manna-grass (*Puccinellia maritima*) seems to be a particularly effective colonist of old pans, and it is often associated with annual glasswort (*Salicornia stricta*) or annual seablite (*Suaeda maritima*). More usually, the sward of vegetation between neighbouring pans is cut through by the erosion of their edges by the currents at high-tide, so that two or more of them join together to form what is called a "compound pan." Many of those appearing in the early phases are transitory, but where a thick layer of vegetation becomes well established round them, they often remain as a permanent feature. All depressions originating in this way are known as "primary pans."

The other method by which they may be produced is by the obstruction of a creek. This may occur either by the extension of the vegetation along its edges to form a bridge of plants which collects the silt and stops up the exit, or by the caving in of its sides as a result of undercutting by the water flowing through it. In the latter case the dam which is produced is usually quickly consolidated by plants establishing themselves over it. The pans formed in this way, known as "channel-pans," are generally long and narrow, and often occur in series, showing the original course of the channel. Sometimes the roofing-in of a creek by the plants along its sides is not sufficient to prevent the flow of water completely, with the result that a series of subterranean tunnels is formed, opening to the surface at intervals. When these occur amongst thick vegetation, they may be quite invisible, but are readily audible when water is running through them. As further silt is deposited on the vegetation, however, sooner or later the tunnels will cave in, and a normal channel-pan will be produced.

In the case of a marsh growing outwards from the land, there will eventually come a time when the outward (horizontal) growth ceases, either as a result of wave-erosion at its seaward edge, or by the scouring action of tidal currents. When this occurs a miniature cliff is sometimes formed along its outer edge, where chunks of the old surface have broken off and been washed away. As a rule this destructive process does not continue for long, unless some permanent change in the prevailing currents has taken place. Usually colonisation soon starts again on the material which has been washed from the base of the little cliff and a fresh marsh begins to form at a somewhat lower level. This type of development is known as "secondary marsh," and whenever one comes across a sudden change in the level of the surface (usually not more than a foot in height), appearing as a continuous line, it probably owes its origin to this process. The secondary marsh frequently shows a lumpy surface, particularly near its junction with the original portion of the marsh. This is due to its having developed on top of the blocks of material which fell from the higher marsh, and as a result there are many hollows which easily become pans ("secondary pans"), similar in appearance to those already described. Small-scale development of secondary marsh can be well seen in many west coast salt-marshes, where the channel banks have subsided as a result of undercutting by the strong flow of the tide, and the new surface has been quickly invaded by plants from the lower levels.

From the above summarised account of the way in which salt-marshes grow up it will be clear that they provide an admirable study of progressive colonisation, and illustrate how the earlier communities, by altering the nature of the habitat, make it possible for later communities to become established (see p. 28). In most salt-marshes it is possible to see a number of these phases in the succession arranged in a series of well-marked zones corresponding to the different levels of the surface.

RELATION BETWEEN VEGETATION AND THE PERIOD OF SUBMERGENCE

The main factor in determining the succession, as we have already seen, is the rise in level caused by the deposition of solids by the tidal waters, which has the effect of differentiating the zones according to the length of time they are exposed or submerged. The actual length of these periods has been given considerable study, and has been found to vary widely over the various levels in a salt-marsh. Thus in the lowest zone of all, commonly inhabited solely by algae or eel-grass (*Zostera* spp.) and well below the level of what is usually looked upon as the salt-marsh proper, it was found in one area that the average time during which it was submerged per month (of 732 hours) was

282 hours. The period of submergence was found to become steadily shorter as the level of each zone increased until in the sea-rush zone, at the highest point on the marsh, it averaged only a mere 3 hours in the month. These results make it clear that there is no part of a salt-marsh, which is capable of supporting plants, that is submerged for even half the available time and that the upper levels are inundated for only very short periods each month. It is perhaps worth while to point out that the periods when a marsh is likely to be submerged for the longest time will occur either at the spring equinox, when the young seedlings are just coming up, or at the autumnal equinox, when many plants are in fruit. The high tides during the autumn are particularly important, since the distribution of the seed of many halophytes is brought about entirely by water.

RATE OF ACCRETION

Some interesting measurements have been carried out in various salt-marshes on the rate at which the silt is deposited. It is obvious that this must depend on a number of factors and is likely to vary considerably from place to place, and also in different parts of the same marsh. Some of the principal factors concerned are the local strength of the tides, the amount of solid matter carried by the water and whether this is derived from a muddy river or from the shore, and the average density of the vegetation. Generally speaking, the results show that sedimentation is most rapid in the lowest levels of a marsh, if we exclude the open communities of pioneer plants along the growing edge, which do not occur sufficiently thickly to trap the solids efficiently. In addition, it is usually found to be greater nearer the larger creeks, provided that the vegetation is well-developed along their banks. In the middle and upper levels, the rate of deposition was found to slacken off gradually as the tidal submergences became less frequent, as would be expected.

From experiments carried out on the marshes near Scolt Head (Norfolk) by V. J. Chapman, it was tentatively estimated that the rise in level in the open community of annual glasswort (*Salicornietum*) along the lowest edge of the marsh was taking place at the rate of 0.68 cm. per annum. In the much thicker vegetation provided by the zone of sea-aster (*Asteretum*), which occurred immediately above this, the rate increased by nearly 50 per cent to 0.98 cm. per annum, but in the zones dominated by thrift (*Armerietum*) and sea-plantain (*Plantaginetum*), higher up the marsh, it was found to fall to about 0.40 cm. It is interesting to use these figures to make a rough estimate of the time it would take the various areas of a salt-marsh to develop from the bare mud. The following table, which gives the estimated times for the establishment of the main zones, suggests that in these marshes it would take about 200 years to reach the level dominated by the sea-rush (*Juncetum*) starting from the stage when the annual glasswort had just become established as a very open community.

VEGETATION ZONE	AVERAGE MAXIMUM DEPTH OF SILT AT CONCLUSION OF PHASE	AVERAGE TIME TAKEN FOR ACCUMULATION OF THIS DEPTH OF SILT
Glasswort (*Salicornietum*)	1.3 feet	58 years
Sea-aster (*Asteretum*)	2.2 feet	106 years
Sea-lavender (*Limonietum*)	3.2 feet	121 years
Thrift (*Armerietum*)	3.3 feet	172 years
Sea-plantain (*Plantaginetum*)	4.2 feet	201 years

Similar experiments carried out by F. J. Richards round the estuary of the Dovey, on the west coast of Wales, produced results which varied considerably according to the position relative to the river, but seem in general to suggest a more rapid increase in height. Here the sediment contained much more sand than at Scolt Head, but there was a specially rapid deposition of silt near the head of the estuary where the muddy river water meets the salt water from the sea.

If the upward growth of a salt-marsh is as rapid as is suggested by these experiments, it follows that its spread horizontally should usually be even quicker. An examination of old maps showing the extent of various salt-marsh areas in the past leaves no doubt that this is true, and that new land is being steadily gained at many places round the coast.

THE SALT FACTOR

We have already seen (page 38) how halophytes are especially equipped to deal with the salinity of the soil-water, which is the master-factor controlling the whole vegetation in salt-marshes. It will be remembered that not only are they able to exert much greater osmotic pressures than ordinary plants, but they can also alter this

R. P. Bagnall-Oakeley

Plate XIII   Open community of annual glasswort, *Salicornia stricta*, colonising mud.   Scolt Head, Norfolk.

a

R. P. Bagnall-Oakeley

b

R. P. Bagnall-Oakeley

Plate XIVa Rice-grass, *Spartina townsendii*, colonising mud; in distance, Blakeney shingle spit. Cley, Norfolk.

 b Mature " Spartina-meadow " at Blakeney, Norfolk (1950), originating from six plants established by Prof. F. W. Oliver in 1929.

pressure in response to changes in the salt-concentration of the soil-water. A good many measurements of the salt contained in the soil-water in different parts of typical salt-marshes have been carried out, and these results leave no doubt that it varies over a wide range throughout the year at all levels. The amount of salt present will clearly depend on the number of times the area is submerged during a given time and on the amount of rain falling upon it in the same period. It has been shown that the widest variations in the salt-content occur in the central portions of most marshes. These areas are subject to relatively infrequent flooding and, after a long spell of dry weather, may acquire a very high salt-content. Prolonged wet weather naturally has exactly the reverse effect. It is a well-known sight in dry summers to find the main vegetation much more stunted than it is in a wet year. The lower levels, which are submerged more frequently, never show such high salt-concentrations in dry weather, nor such low ones in wet weather, since they are continually exposed to salt water of constant strength. The amount of salt is also less in the highest levels, since these are very rarely submerged by the tide and the rainfall is normally sufficient to keep it down to a small amount.

It has been shown that the maximum concentration of salt reached during dry periods in the main central portion of a marsh remains roughly the same, no matter which of the different zones of " middle level " vegetation is considered. A value of 6 to 7 per cent (expressed as per cent of the water present in the soil) is not unusual, and in the practically bare soil of the salt-pans it may become as high as 16 per cent on the surface, although a few inches deeper it will probably not exceed 2 to 3 per cent. It is clear that the concentration of salt in the surface-layers of the pans is too high for the germination of any plants, with the possible exception of the glassworts (*Salicornia*), and this, coupled with the absence of drainage, accounts for their general bareness. (For purposes of comparison it should be remembered that sea-water normally contains about 3 per cent of salt.) It is interesting to note that the salt-values at all levels fall to a minimum in the early spring, as a result of the winter rains and the slower rate of evaporation during the colder weather. This is, of course, the time when the germination of seeds is likely to be taking place most actively.

It may seem surprising that quite a number of different halophytes should be found happily established and obviously flourishing in the middle levels of most marshes, where the salinity has been shown to be most variable. The reason is that most of these plants are perennials equipped with deep roots, which enable them to draw their water supplies from regions where the salt-content remains reasonably constant and never reaches the high values sometimes found near the surface. Moreover, when a certain number of these plants have established themselves, they commonly spread vegetatively rather than by seeding. We shall have occasion to refer to the relation between the plants found at different levels and the salt-content later on in this chapter (page 81), after the vegetation of the various zones has been described in detail.

THE EFFECT OF TIDAL CURRENTS ON PIONEER SPECIES

A high concentration of salt is not the only difficulty with which the earlier colonists have to contend. The various annual species of glasswort, which are common pioneers on bare mud, suffer heavily from the effects of strong tidal currents, which dislodge large numbers of seedlings before they have had time to become firmly established. In some experiments carried out by P. O. Weihe on a colony of *Salicornia stricta*, it was estimated that not more than 25 per cent of the plants which germinated survived in the " neap-tide zone " (i.e. where they were submerged regularly every 12 hours). In the " spring-tide zone," however, where the period of submergence varied from 1 to 15 days, the average number that survived worked out at 65 per cent. It was also noticed that a dense community of glasswort was never established unless it was in a position where the young seedlings could expect at least three clear days undisturbed by the tide. When the plants had once become properly established on the mud, a daily submergence did not appear to have any adverse effect on their growth. These plants, since they are annuals, never develop more than shallow roots (Fig. 9(a) p.49), and so even fully-developed specimens are dragged from their moorings from time to time, if the currents are strong.

Having discussed some of the main characteristics of salt-marshes, we are now in a position to consider in detail the composition of the vegetation with which they are covered. A great deal of field-work has been carried out on this habitat, indeed far more than on any of the others with which we are concerned. Observations from a large number of different localities make it clear that, though the same kind of zonation can be seen in most marshes, the actual succession varies a good deal from place to place and certain individual species appear earlier in some marshes than in others. Generally speaking, the most complex development is to be found on the Norfolk marshes, those along the south and west coasts showing a simpler series of stages. These differences probably depend on a number of factors, but perhaps the most important is the nature of the soil-material. Thus most of the west coast marshes are distinctly sandy, the East Anglian ones usually of fairly firm mud, while those along the south coast are formed on soft sloppy mud. In the survey which follows, a selection of the most typical zones of vegetation found in various areas is described in order of ascending height. It is hardly necessary to add that no given marsh will show *all* these zones, though in point of fact examples of the majority of them can be found somewhere among the East Anglian marshes. A summary of typical successions for different parts of the country is given at the end of this chapter (page 83).

LOWEST ZONE: COMMUNITIES OF ALGAE OR EEL-GRASS

The lowest zone of vegetation is usually found to begin just below the low-water mark of the neap tides. This generally consists of an open community of algae (seaweeds), which are common pioneers on mobile mud or sand. It is usually made up of green algae such as *Enteromorpha*, *Vaucheria* or *Rhizoclonium*, and almost pure communities of one or other of these may sometimes be found. These green sea-weeds are familiar objects all along the coast, particularly on rocks, harbour piers, in shallow pools and brackish water generally. If they are exposed to the air for long periods they are often bleached white. Algae occur at all levels of a salt-marsh, but their zoning does not appear to be clearly correlated with the level at which they are found. It can usually be noticed, however, that the algal population is much richer in muddy than in sandy marshes.

In some places, a quite different community occurs at the same level, dominated by the eel-grass or grass-wrack (*Zostera* spp.). These plants are perennials with long grass-like leaves, and may form extensive submarine meadows, extending well below the low-tide level. The communities are largely pure (*Zosteretum*), but may be associated with marine algae and sometimes with the inconspicuous tassel-pondweed (*Ruppia maritima*), a much smaller plant with very narrow, alternate leaves.

These earliest colonists undoubtedly play a not inconsiderable part in trapping silt and stabilising the ground before the arrival of other plants. In addition, their decaying remains add a certain amount of humus, which improves the soil.

ANNUAL GLASSWORT ZONE (*Salicornietum*)

The first well-marked community on the salt-marsh proper (i.e. the part which is uncovered for much of the day) in most places is a consociation dominated by annual glasswort or marsh samphire (*Salicornietum*) (Pl. XIII, p. 66). A number of closely allied annual species of *Salicornia* have been described, but they are difficult to separate and we can take the common *Salicornia stricta* (*herbacea*) as typical. It is an ugly little plant, with erect, highly succulent stems from 6 in. to a foot in height. These are made up of a number of jointed segments, which carry no proper leaves, the leaf-bases alone existing in the form of scales (Fig. 9(b) p. 49). This community is usually very open, showing plenty of bare mud between individual plants, and often interspersed with various algae. This is particularly true of its lowest edge where, as we have seen, many young plants may be uprooted by the strength of the tidal currents. Though these plants are only annuals, their dead stems generally remain stiffly erect for some time and are effective in encouraging the deposition of silt as the tide flows by. Sometimes the surface of the mud between the plants becomes covered with quite a thick mat of algae, which render it far more stable and greatly improve the chances of successful germination of the seeds. If one moves inwards from the open edge of the *Salicornietum*, it will be noticed that the vegetation becomes gradually thicker, and that certain other halophytes begin to appear as associates. The first arrivals are usually annual seablite (*Suaeda maritima*) (Pl. 3, p. 71), sea-aster (*Aster tripolium*), and sometimes sea manna-grass (*Puccinellia maritima*). The last-named sometimes appears as a pioneer instead of glasswort, particularly in some west coast marshes, or it may elsewhere share the dominance of the first community with glasswort. On the majority of the muddy East Anglian marshes, however, it appears much later in the succession (see page 84).

John Markham

Plate 3 Annual Seablite, *Suaeda maritima*; a common salt-marsh plant, shown in its autumn colouring

Rice-Grass Zone (*Spartinetum*)

A quite different pioneer colonist, particularly characteristic of the deep mobile mud of the south coast, but also established locally elsewhere, is rice-grass (*Spartina townsendii*). This remarkable plant is an exceptionally strong-growing perennial, thriving best in deep, soft mud which is too unstable for the successful establishment of annual glasswort. It appears to be a hybrid formed by the natural doubling of chromosomes between the cord-grass (*Spartina maritima* (*stricta*)), a native plant found locally in a number of marshes in similar positions, and *Spartina alterniflora*, a rare alien grass from North America, first noticed on the south coast in 1829. The hybrid was reported from Southampton Water in 1870, and since then has spread with extraordinary rapidity along the south coast. It is now established also at a number of places on the east coast, in the Severn estuary, and even as far north as the Mersey. In addition, it has been planted in various localities in connection with land-reclamation, and may perhaps become a familiar feature on most of our salt-marshes in the future.

It is a much larger plant than our native *Spartina maritima*, growing in tussocks 1 to 4 feet high, with rather narrow leaves standing out at an angle to its stem and producing tall cylindrical flower spikes. It has extensive creeping rhizomes, which bind the soft mud very efficiently. Its roots are of two kinds; a few long anchoring ones growing vertically downwards, which hold the plant firmly in the mud, and a large number of fine absorbing roots, which grow out horizontally just below the surface (Fig. 9 (c) p. 49). The erect habit of the plant makes it extremely effective in slowing down the rate of flow of the tides, so that large amounts of silt and vegetable flotsam, in the form of seaweeds and plant-remains, are caught and held round its shoots after the water has receded. Its thick tussocks are also instrumental in restraining the removal of mud by the strong ebb-tides. Rice-grass outstrips all other salt-marsh plants in its powers of collecting silt from sea-water, so that the rate at which the level of the mud-flats rises is much higher in the *Spartinetum* than in other communities. The remains of seaweeds that collect round the tussocks, as well as the dead leaves of the plants themselves, add humus rapidly to the soil, and in a very few years the bare unstable mud may become transformed into a thick sward, which can provide satisfactory grazing. Its value in gaining land from the sea is obvious, and it has been planted in a number of places for this purpose.

As a rule rice-grass is found as a largely pure community, with perhaps occasional plants of sea-aster or annual glasswort amongst it, but locally sea-purslane (*Halimione* (*Obione*) *portulacoides*) is sometimes co-dominant in the lower marshes. In its early stages it appears as a number of isolated clumps surrounded by bare mud (Pl. XIVa, p. 67), but it rapidly spreads over the intervening spaces to produce a uniform carpet. This mature *Spartinetum*, when it is used for grazing, is sometimes known as " Spartina meadow " (Pl. XIVb, p. 67). In the extensive marshes in Hampshire and west Sussex where rice-grass is now the primary colonist, *Salicornietum* covers only very small areas or may be completely absent. Probably in the past the native *Spartina maritima* occupied roughly the same position in these south coast marshes as its more virile hybrid does to-day, since the mud must always have been too unstable for the successful establishment of a *Salicornietum*, but there is no evidence that it ever covered the same large areas of the middle levels. If in the future rice-grass should spread widely to other parts of our coasts, it may oust the annual glasswort from its present position as a marsh-builder and even spread over the higher zones as well. It seems likely, however, that it requires deep soft mud for its most vigorous development, and these conditions occur only locally, chiefly in wide estuaries, apart from the main south coast areas where it is already firmly established. For this reason a general invasion of all our salt-marshes in the future is fortunately unlikely, for it would destroy much of their interest.

SEA MANNA-GRASS ZONE (*Puccinellietum*)

Returning now to those marshes where *Salicornietum* occurs typically, we have already mentioned that the upper edge of this zone is usually invaded by other plants, and one or other of these later arrivals eventually gains dominance in the next consociation. Taking the country as a whole, sea manna-grass (*Puccinellia* (*Glyceria*) *maritima*) is the most usual successor, for by extending rapidly in a horizontal direction to form a turf, it soon replaces the vertically-growing glasswort as a dominant species, though it does not choke it out completely. It usually forms a perfectly distinct zone in the form of a fairly close sward, sometimes with a rather hummocky surface, immediately above the *Salicornietum*. With its many branching stems, it offers a much greater obstacle to the tides, and collects silt rapidly. It undoubtedly prefers a sandy substratum, for in the muddy east coast marshes it arrives at a later stage in the succession and then rarely spreads widely. In the West, on the other hand, it is quite often the first colonist, or it may appear simultaneously with annual glasswort to form a mixed community in which the two species are co-dominants. Typical *Puccinellietum* (formerly known as *Glycerietum*) commonly contains a number of associates, but these vary considerably from place to place, and it sometimes occurs as an almost pure community. Large areas of salt-marsh vegetation, particularly in the West, are either dominated by sea manna-grass or are jointly dominated by it and some other halophyte. They are extensively used for grazing. The following list gives some of the typical associates found in this consociation in various localities:

scurvy-grass	*Cochlearia officinalis* (Pl. XXXII, p. 167)
sea-spurrey	*Spergularia marginata*
sea-aster	*Aster tripolium*
sea-plantain	*Plantago maritima* (Pl. XXXVII, p. 198)
annual glasswort species ..	*Salicornia stricta*, etc.
annual seablite	*Suaeda maritima*
sea-arrowgrass	*Triglochin maritima* (Pl. XXXIX, p. 206)

SEA-ASTER ZONE (*Asteretum*)

In other marshes, particularly those on the east coast, sea-aster is the first invader of the *Salicornietum*. This is one of the best known halophytes and a distinctly handsome " daisy." It is a large plant, reaching a height of at least two feet, and it soon displaces the glasswort from its dominance. There is a variety of this plant (var: *discoideus*) (Pl. 4, p. 78), lacking the purple florets round its golden discs, which is particularly characteristic of the East Anglian marshes, but is rarely found north of the Humber estuary. In the early stages of the development of *Asteretum*, annual glasswort or seablite may be equally abundant, and the latter at any rate usually remains a common associate in this zone at all levels. In some places, e.g. along the Wash, *Spartina maritima* may be locally dominant over small areas, and precedes the establishment of *Asteretum*. When well-developed, the sea-aster zone appears as a thick and comparatively tall sward, which encourages the rapid deposition of silt when it is covered by the tide. Any of the plants listed as associates in the *Puccinellietum* may also occur in this consociation, and there is often a thick layer of algae as well. When the sea-aster is in flower in late summer, this zone makes a bright sight, though hardly comparable to the brilliance of a sheet of sea-lavender or thrift.

We have taken the development of either *Puccinellietum* or *Asteretum* at this level as examples of the typical succession to *Salicornietum*, but sometimes sea manna-grass and sea-aster invade it simultaneously, and all three species may be equally abundant for a time, though as a rule the close turf produced by the former will eventually secure it the dominance.

Much of the area of a typical salt-marsh which we have so far considered belongs to that portion sometimes called " low salt-marsh," the greater part of which is covered by average high-tides. All the subsequent zones, which are only inundated by the higher spring-tides, are known as " high salt-marsh."

SEA-LAVENDER ZONE (*Limonietum*)

Just above the level we have been discussing, a zone may sometimes be found in which the common sea-lavender (*Limonium vulgare*) is dominant, or co-dominant with either sea-aster or sea manna-grass. Its presence is by no means universal, but it may be looked upon as a typical community of the middle levels, and where it does occur, often covers considerable areas. In late summer, when the plant is in full bloom, this zone shows up in the distance as a glorious sheet of mauve and is perhaps the most lovely sight that a salt-marsh can offer (Pl. 5, p. 87). Occasionally the rarer " remote-flowered sea-lavender " (*Limonium humile*) can be found growing with the common species, and in Ireland this is the usual species seen. The following list of associated species found in the *Limonietum* in various marshes shows that they do not differ much from those occurring at slightly lower levels:

scurvy-grass species	..	..	*Cochlearia officinalis* and *C. anglica*
sea-spurrey species	..	..	*Spergularia marginata* and *S. salina*
sea-aster	..	..	*Aster tripolium* (often co-dominant)

remote-flowered sea-lavender	*Limonium humile* (*rariflora*)	
thrift	*Armeria maritima* (Pl. II, p. 7)	
sea-plantain	*Plantago maritima*	
annual seablite	*Suaeda maritima*	
sea-arrowgrass	*Triglochin maritima*	
sea manna-grass	*Puccinellia maritima* (often co-dominant)	

The absence of a *Limonietum* zone in some marshes, particularly those in the West, may possibly be due to heavy grazing.

THRIFT ZONE (*Armerietum*)

Sometimes a zone will be found just above the sea-lavender, or at any rate somewhere on the central or upper levels of a marsh, which is dominated by thrift (*Armeria maritima*). Thrift is probably less tolerant of salt than many halophytes, and it always flourishes best when the salinity is low. Owing to its very deep roots, however, it is able to avoid the rapid changes of salt-concentration in the surface-layers, which are typical of this part of the marsh. The *Armerietum* usually takes the form of a close turf, sometimes with sea manna-grass as a co-dominant. It stands up to grazing very well, adopting a compact rosette under these conditions, very different from its more diffuse habit when growing in partial shade and with a good water supply (Fig. 6), p. 45. The associate species found in this zone are much the same as those listed above for the *Limonietum*. A large strip of *Armerietum* when the thrift is in full flower is a glorious sight. It is usually at its best towards the end of May, but its flowering season is considerably longer than that of the sea-lavender. If you prefer pink to purple, you may decide that the thrift is the more beautiful.

THE "GENERAL SALT-MARSH COMMUNITY"

On many marshes neither of the above consociations is found in the form of a definite zone, though both dominant species are present. Instead, the vegetation of the middle levels is found to be made up of a number of different plants, none of which appears to be generally dominant. When such a zone occurs, it is usually known as the "general salt-marsh community," or just as "saltings." Thus at Scolt Head, the following six plants form a community in which they all appear to be co-dominant:

sea-spurrey	..	..	..	*Spergularia salina*
sea-lavender	..	..	..	*Limonium vulgare*
thrift	..	..	..	*Armeria maritima*
sea-purslane	..	..	..	*Halimione (Obione) portulacoides* (Pl. 16 p. 210)
sea-arrowgrass		..	..	*Triglochin maritima*
sea manna-grass		..	..	*Puccinellia maritima*

It is distinctly unusual to find six equally dominant plants in any type of vegetation, and a possible explanation in this case is that their root-systems reach to different depths and their main flowering seasons occur at different times. The " general salt-marsh community " is found at the same general level as the *Limonietum* or *Armerietum* on other marshes.

SEA-PURSLANE COMMUNITY (*Halimionetum*)
(formerly known as *Obionetum*)

Sea-purslane (*Halimione portulacoides*) plays a rather special part in some salt-marshes, particularly those on the east coast. It is a bushy little plant, whose long stems lie mostly along the ground. Its greyish-white leaves show up prominently in the distance, and it often forms a very characteristic society along the banks of creeks. It can be found in this position at all levels except the very lowest, and can even survive in the thick vegetation of pure *Asteretum*. It appears to prefer a well-drained soil, in rather the same way as does the shrubby seablite (*Suaeda fruticosa*) when growing in shingle (see page 128), and the sides of these channels offer such conditions. This comes about in the following way. As the lower creeks fill up at high-tide, there comes a moment when the water overflows across their edges on to the adjacent marsh. Any plants fringing the sides will filter off and retain a portion of the solids carried by the water, with the result that the level of the bank will be gradually raised above that of the surrounding marsh. As this position becomes better drained, the sea-purslane grows more and more luxuriantly, and eventually swamps any other vegetation. A bird's-eye view of an east coast marsh would show nearly every little creek bordered with a light grey band produced by the presence of these plants.

Sometimes it spreads like a weed from this position right over the central levels of a marsh, if they are reasonably dry (Pl. XV, p. 82). When it does this, it soon forms large patches which completely obliterate the previous vegetation. The only species, in fact, which seems to be able to survive in really thick *Halimionetum* is sea manna-grass, which forms a lower layer of vegetation and is therefore not an active competitor for space. Sea-purslane also flourishes if there is some blown sand on the surface of the marsh, as may well happen if there are sand-dunes in the vicinity. When growing under these conditions it often forms tussocks, or miniature dunes, as a result of sand collecting round the plants. Although sea-purslane is abundant along much of the east coast, it is more local and largely confined to the creek-sides on the south coast, and generally rather rare in the West.

A purely local peculiarity of certain Norfolk marshes is the occurrence in these relatively mature areas of certain unusual plants which are more characteristic of shingle. They do in fact occur also on shingle and shingly-sand in the neighbourhood (see page 132). All these are typical Mediterranean species, and they grow here in company with sea-purslane and other typical plants of the middle levels. The following is a list:

sea-heath	.. .. ..	*Frankenia laevis*
matted sea-lavender	.. ..	*Limonium bellidifolium* (*reticulatum*)
shrubby seablite	.. ..	*Suaeda fruticosa* (Pl. XXXIII, p. 114)

RED FESCUE ZONE (*Festucetum rubrae*)

On sandy and well-drained marshes, especially those on the west coast, red fescue (*Festuca rubra*) often becomes established at a level above the " general salt-marsh community " or comparable zones, and may become dominant. It sometimes starts as a co-dominant in the upper *Puccinellietum*, but eventually it usually gains the dominance. At its upper edge, on the other hand, it is often co-dominant with the mud-rush (*Juncus gerardi*) for a time, until it has to give way before the taller plant. This region of a marsh is but rarely flooded, and is apt to become very dry during settled weather in the summer, but this grass stands up to desiccation well. Red fescue is found at a much lower level in some marshes, and may even be co-dominant with annual glasswort and seablite, but the most characteristic position for *Festucetum rubrae* is on the higher levels, where it often covers wide areas. Almost any of the usual salt-marsh halophytes may be found in this consociation, but those in the following list are perhaps the most typical:

sea-spurrey	*Spergularia marginata*
thrift	*Armeria maritima*
sea-milkwort	*Glaux maritima*
buck's-horn plantain	*Plantago coronopus* (Pl XXXVI, p. 179)
sea-plantain	*Plantago maritima*
sea-arrowgrass	*Triglochin maritima*
sea hard-grass	*Parapholis strigosa* (formerly *Lepturus filiformis*)

SEA-PLANTAIN ZONE (*Plantaginetum*)

In certain east coast marshes a zone can sometimes be distinguished at this level in which the sea-plantain is dominant (*Plantaginetum*). This forms a close turf in the same way as thrift, and, like the *Festucetum*, is generally in the grazing area. The associated species are much the same as those found in the " general salt-marsh community," and this consociation may be taken as a typical example of various minor communities found in East Anglian marshes, which are really offshoots from the main line of the succession.

SEA-RUSH ZONE (*Juncetum*)

The highest zone in many salt-marshes takes the form of community dominated either by the sea-rush (*Juncus maritimus*) or the mud-rush (*Juncus gerardi*). When present, it is always a conspicuous feature on account of its much greater height in comparison with the low close turf in the flats below. The presence of a tall plant as dominant in these communities has the effect of destroying the close sward characteristic of the middle levels and a certain amount of open soil appears once more. This soil is, moreover, moister owing to its being partially shaded, and for this reason certain plants typical of the more open and damp lower levels sometimes reappear in this zone. Sea-aster, scurvy-grass and even annual glasswort may well be seen here in some quantity, and there is usually an increase in the number of algae. Occasional inland plants, too, may turn up in the

Plate 4   Ray-less form of sea-aster, *Aster tripolium* var : *discoideus*

Juncetum, since this area is rarely flooded. It is unnecessary to give a list of the typical associates in this consociation, since practically any of the salt-marsh halophytes mentioned in this chapter may appear here. From my own impression, I should say that sea-milkwort is particularly characteristic, and that amongst the inland species, white bent-grass (*Agrostis stolonifera*) is a frequent inhabitant.

Many salt-marshes do not develop a *Juncetum* zone at all. This is sometimes due to their enclosure by sea-walls or to artificial draining. Grazing, on the other hand, appears to have no effect on it, the rushes being left severely alone by cattle. Where there is no belt of rushes, the upper edge of the marsh is often marked by a fringe of vegetation consisting of the sharp couch-grass (*Agropyron pungens*), a similar plant to the sea couch-grass, but with rather broader and rougher leaves, sea-wormwood (*Artemisia maritima*) (Pl. XXXI, p. 166) and sometimes sea-barley or squirrel-tail grass (*Hordeum marinum*). If there is any sort of bank to keep out the sea, it is likely that these plants will be found on it. They may also occur in the *Juncetum*, sea-wormwood in particular being commonly found in the upper levels of many marshes bordering the sides of drainage channels, where its white downy leaves make it conspicuous.

TRANSITION TO FRESH-WATER SWAMP

The *Juncetum* may be looked upon as the final stage in the development of the salt-marsh proper (i.e. where the vegetation is still influenced by the master-factor of a saline soil). We have, however, referred already to the fact that in some places the *Juncetum* may be gradually replaced by a fresh-water swamp, particularly where a supply of fresh water comes in from the land. In such cases a community grows up with the common reed (*Phragmites communis*) as the dominant species and perhaps one of the sea club-rushes (*Scirpus maritimus* or *S. tabernaemontani*) as a local sub-dominant. These last are typical inhabitants of brackish water (see page 158); the former having a triangular stem, the latter a cylindrical one and closely resembling the common bulrush. The sea-rushes (*Juncus maritimus* or *Juncus gerardi*) are likely to be still present in this community, but hardly in sufficient quantity to be called dominants. A number of the halophytes usually seen in the *Juncetum* will continue to survive here too, but they are joined by a number of fresh-water marsh plants. Some of the most characteristic non-maritime plants likely to be found in this zone are:

lesser spearwort	.. ..	*Ranunculus flammula*
parsley water-dropwort	..	*Oenanthe lachenalii*
brookweed	.. .. ..	*Samolus valerandi*
rushes (fresh-water)	.. ..	*Juncus* spp.
reed canary-grass	.. ..	*Phalaris arundinacea*

Needless to say, there is no sharp dividing line between the end of the *Juncetum* and the beginning of the *Phragmitetum*, but on the landward side of the latter it is usually quite clear that the influence of the sea is no longer operative.

Possible Succession to Woodland

The natural transition from salt to fresh-water marsh can be seen in several places on the Atlantic coast of North America, and may well have occurred on a considerable scale in this country in the past, particularly in the Fenlands. There is no reason why, in the natural course of things, reed-swamp should not be succeeded in coastal districts by what is called " carr " (swamp-woodland or scrub), as can still be seen taking place in the Fens. If this happened, the carr would eventually be replaced by woodland, the climatic climax for the country. In the British Isles, however, agricultural operations have removed all traces of such a succession having occurred in the past, and the theoretically possible development of forest from a salt-marsh cannot be observed anywhere.

There is actually some doubt as to whether, in the absence of fresh water, a completely inland vegetation can ever be developed from a salt-marsh, unless the area is artificially segregated from occasional flooding by sea-water by the construction of some sort of barrier. In this connection there is some evidence that the land north of a line from near the Humber to the Mersey is slowly rising in relation to the sea-level, whereas the land to the South of it is falling. If this is a fact, it is clear that in any locality where the land level is rising, sooner or later the halophytic vegetation must give way to normal inland vegetation, and that no artificial aid is required. The vegetation round the Solway Firth is thought by some botanists to show clear

signs of this natural change taking place. In those places where the
land-level is falling, it is equally clear that the salt-marshes will
gradually disappear under the sea, unless the rate at which the solids
are being deposited can keep pace. Obviously the natural development
of inland vegetation here is very unlikely.

In the absence of more definite evidence, we must leave open the
question as to whether or not an inland type of vegetation can be
naturally developed. There can be no doubt, however, that when it
takes place it represents a complete change in the conditions of the
habitat since fresh water replaces salt water in the soil, and the most
important factor in determining the previous vegetation is thus
removed. We can therefore look upon the *Juncetum* as a climax, or
at any rate a stable sub-climax under the conditions of a saline habitat.
All the communities leading up to this are part of the " halosere," or
stages of the succession in the Salt-marsh Formation.

The Salt-Tolerance of Different Halophytes

In the survey of the typical zones of vegetation given above, it will
have been noticed that the number of different species is very small.
The lists of plants given in the text are not intended to be complete,
but they do in fact include nearly all those normally found in the
lower and central portions of most marshes. In the upper levels,
where flooding is comparatively rare, a wider variety of plants can
be found, and by no means all of these have been mentioned by name.
All through this chapter the relation of the vegetation to the level at
which it occurs has been purposely stressed, but it is noticeable that
there are some plants such as annual glasswort, annual seablite, sea-
aster and sea manna-grass, that are found to a certain extent at all
levels. Others, like the sea-rush, sea-wormwood or the sea hard-grass,
seem to be confined to the higher levels.

Observations carried out on the Scolt Head marshes by V. J.
Chapman, involving reasonably accurate estimations of the frequencies
of the main species at all levels, suggest that for many plants there
is a general relation between their frequency in a particular position
on the marsh and the number of times they are submerged by sea-
water per annum. These quantitative results make it clear that the
plants found most widely at all levels nevertheless reach their greatest
abundance at the lower levels, no doubt on account of the absence of

competition. In the higher levels they tend to be crowded out by other halophytes which are unable to tolerate the greater salinities of the lower marshes. There is no evidence that the plants growing in the lower levels actually require a high concentration of salt, but they can clearly tolerate greater salinity and consequently grow more thickly when their competitors are excluded. There are, however, a number of anomalous plants, whose distribution does not fit in with this scheme and must be controlled by other factors not yet understood.

CONVERSION OF SALT-MARSHES TO AGRICULTURAL LAND

The upper levels of many salt-marshes have in the past been fenced off by sea-walls and banks, so that the tides are permanently excluded from them. When this is done, the salt remaining in the soil is soon washed out and the land becomes capable of supporting a much wider variety of inland grasses, eventually losing all the characteristics of a salt-marsh. If a simple drainage system is constructed, excellent " salt-pasture " can be produced, and, if this is subsequently ploughed up and cultivated, good arable land may result. The value of this land will depend essentially on the amount of silt and fine particles which were deposited when the marsh was originally formed. The rather sandy marshes typical of much of the west coast, for instance, do not make such satisfactory pasture when they are reclaimed as those elsewhere containing more silt, and they are rarely suitable for conversion into arable land. Nevertheless, the natural wild marshes of the West furnish a great deal more grazing than those on the east coast, since the area of grass (*Puccinellietum* or *Festucetum*) is usually much larger. On the other hand in the East, especially round the Wash and the south Lincolnshire coast, some of the finest agricultural land in the country has been produced by the reclamation of old areas of salt-marsh. Perhaps scenically this district of rich flat farmland is not very attractive, but none the less there is something rather fascinating about land which has been gained from the sea by utilising a natural process in this way. Anyone familiar with the present appearance of the Wash at low-tide can easily visualise what the fertile agricultural land round it must have looked like in the past, and also what the present useless mud-flats beyond the edge of the marshland might possibly become in the future. At the same time

R. P. Bagnall-Oakeley

Plate XV Sea-purslane, **Halimione portulacoides**, bordering a creek and spreading over the adjacent marsh. In distance, Blakeney shingle spit with clumps of shrubby seablite, *Suaeda fruticosa*. Cley, Norfolk.

Plate XVI Sea-rocket, *Cakile maritima*, a common " strand " plant on sandy shores. Cornwall.

it must be emphasised that the production of fertile soil is often a slow process, and that it is useless to enclose and drain an area of salt-marsh until a sufficiently thick layer of silt has accumulated on the surface by natural means. The substratum of the Wash is generally sandy, and if too large an area of immature marsh were enclosed, the resulting soil would have little agricultural value.

TYPES OF SUCCESSION

In the course of our survey of typical salt-marshes it has been necessary to generalise considerably in an attempt to give a picture of representative stages by which the mature vegetation of the upper levels is reached. All marshes show some sort of succession, but the composition of the various zones of vegetation and the order in which they occur varies widely from place to place. Indeed few salt-marshes display exactly the same type of development, and in larger areas one can sometimes recognise several alternative routes leading towards the same final vegetation. An examination of the succession taking place in a number of different marshes shows that it conforms to certain broad geographical types in the British Isles, and we can conveniently conclude this chapter by considering the typical routes by which the marshes on the east, west and south coasts have been developed. Much of what follows is based on the classic work of V. J. Chapman, to whose fine series of papers (see Bibliography) I am indebted for much other information in this chapter.

As a broad generalisation for the succession in most salt-marshes in this country, we can construct the following scheme:

ALGAL COMMUNITIES → LOW MARSH → HIGH MARSH→HIGHEST MARSH → SALT
                Salicornietum   " General   Rush Zone   PASTURE
            Early *Asteretum* salt-marsh "   (*Juncetum*) (reclaimed)
                  etc.   etc.

When we come to details, it is clear that, amongst a number of different factors at work, the main one responsible for differentiating the course of development of one marsh from another is the proportion of sand to mud deposited by the tides. Taking the country as a whole, it can be said that the muddiest salt-marshes are those found on the south and south-east coasts, and that the farther north one travels up both the east and west coasts, the more sandy they become. The west coast marshes are, however, more sandy as a whole than those on the east coast, and those in East Anglia, although muddy, are formed of firm mud or clay, very different from the sloppy mud typical of the south coast. As an example of the way in which the distribution of a species can be correlated with this factor, we may consider that of the sea manna-grass. It has already been mentioned (page 73) that this plant appears to be peculiarly sensitive to small differences in soil conditions, and shows a marked preference for well-drained ground. Thus in the muddy East Anglian marshes it generally plays but a small part, and is never a primary colonist except in one or two small areas where the soil happens to be sandy. In most of the south coast marshes also it plays a quite subsidiary role, but on the sandier north-east coast (e.g. at the mouth of the Tees) it becomes established soon after the annual glasswort, and it may even be a primary colonist in certain sandy areas along the Wash. On the sandy west coast marshes it is the dominant species at nearly all levels, and is much the commonest plant seen. This species is obviously a particularly clear example, but the unequal distribution of many other plants may well be due to similar causes.

The succession on the east coast, particularly on the much-studied East Anglian marshes, is more complex than that found anywhere else. Not only is there a larger number of recognisable consociations, but most of these contain a wider variety of associated species. Another usual characteristic of these marshes is an extensive area of " general salt-marsh " communities near their centre, and an unusually rich algal flora at all levels. Sea-purslane is abundant in all east coast marshes south of the Humber, not confining itself to the sides of creeks as it usually does elsewhere, although nearly always found in greater profusion along them. A complete diagram showing all the possible lines of succession for these marshes would be very complicated, since endless variations are possible, but the following gives some of the main successions usually seen. " Short-cuts " and minor alterations in the order of the communities may easily be produced by local differences in the conditions. The right-hand side of the diagram gives the typical East Anglian succession.

In the salt-marshes along the south coast the main characteristics are the very soft mud and the enormous areas which are covered by the growth of rice-grass. The number of different plants to be seen is very much smaller than on the east coast, and the algal population is generally sparse. No doubt the thickness of the vegetation produced by the *Spartina* is the principal reason for this. Owing to the relatively recent arrival of this grass in all these areas, it is somewhat difficult to work out the order of the succession, since a good deal of the relict vegetation dates from the days before it appeared. It has already been suggested that in times gone by the native *Spartina maritima* may have

EAST COAST SUCCESSION

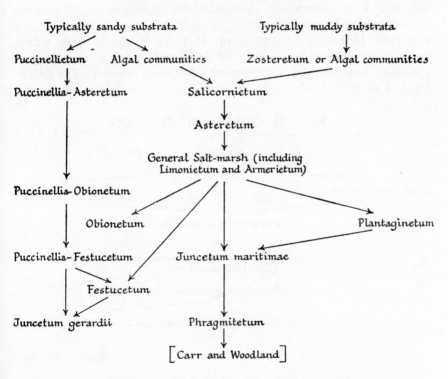

played the same role of primary colonist as its more vigorous hybrid does now, but there is no evidence that it ever spread over large areas, and in any case its effect in swamping other plants could not have been so marked. *Salicornietum* is rarely developed owing to the instability of the mud, but an open community of algae is sometimes present before the ground is colonised by rice-grass. Nowadays, as we have seen, open communities of this grass normally develop into closed

"Spartina meadow," where a few of the usual "middle level" halophytes can often be found, but it is rare to find any considerable area showing what we have called "general salt-marsh" vegetation. Owing to the rapidity with which rice-grass collects silt, these marshes have a particularly grimy appearance in their lower levels, although naturally the leaves of all halophytes subject to frequent submergence become coated with a certain amount of solid matter. When the mature *Spartinetum* is not grazed, it would appear that its normal successor is a community dominated by the mud-rush (*Juncetum gerardi*), which in turn may be replaced by one dominated by the sea club-rush (*Scirpetum maritimae*), or, if there is much fresh water, reed-swamp (*Phragmitetum*) may be developed. Owing to grazing or enclosure, however, this development is not often realised. The following is a simplified scheme for the succession in south coast marshes:

SOUTH COAST SUCCESSION

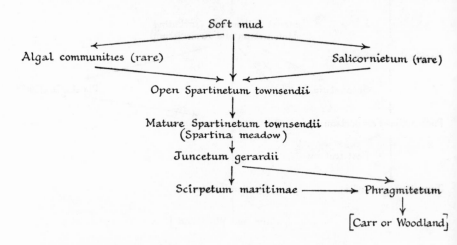

Any reference to the local development of *Puccinellietum*, *Plantaginetum*, *Juncetum maritimae*, etc. in certain south coast marshes has been excluded from the above diagram for the purpose of simplification.

The succession in west coast marshes is probably the simplest. Their most characteristic feature is undoubtedly the dominance of sea manna-grass over large areas as a result of their sandy nature. Although this is sometimes replaced by red fescue on the higher levels,

Eric Hosking
Plate 5   Common sea-lavender, *Limonium vulgare*, in a Suffolk salt-marsh

the general impression of all these marshes is one of grass, and much of this salt-pasture is employed as rough grazing. Sea-milkwort occasionally becomes dominant in the wetter areas, and can generally be counted on to be frequent at most levels. Sea-purslane and sea-lavender, on the other hand, are rather rare, probably because they cannot tolerate the heavy grazing. A " general salt-marsh " com-

WEST COAST SUCCESSION

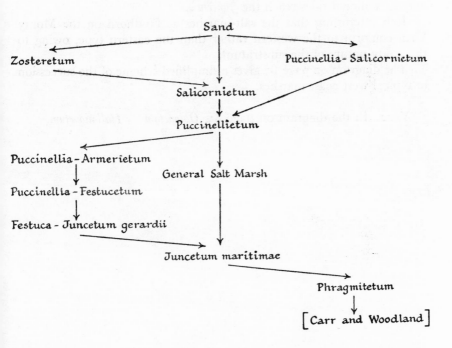

munity is often developed on the middle levels, and it may be said that the number of associated species at all levels is considerably greater than in the south coast marshes, although hardly equal to that on the east coast.

The much-studied salt-marshes round the Dovey estuary in Wales may be taken as a typical example of west coast succession. Here there are five main zones of vegetation:

1. Open *Salicornietum* (not always present).
2. *Puccinellietum* (often the pioneer community).
3. *Armerietum* (often with *Puccinellia maritima* as co-dominant).
4. *Festucetum rubrae* (sometimes with *Juncus gerardi* as co-dominant).
5. *Juncetum maritimae.* In this rush zone, many of the associated plants seen in zones 2 and 3 are more abundant than in the close turf.

The total vertical range in these marshes is just over four feet, and only exceptional tides reach the *Juncetum*.

It is interesting that the salt-marshes at Findhorn on the Moray Firth conform to the western rather than the eastern type, owing to the sandy nature of the substratum.

The diagram on page 87 gives a simplified scheme of the succession in typical west coast marshes.

Note : In the diagram on page 85, *Obionetum* = *Halimionetum*.

CHAPTER 6

STRAND AND FORESHORE
VEGETATION

O N MANY beaches or along the shores of tidal inlets a rather sparse
community of plants may often be noticed growing just above
the usual upper limit of the tide. It may be seen equally frequently
on sand, shingle or firm mud, and in its most typical form appears
in the distance as a definite thin line along the shore. It usually
occurs a short distance above the deposit of dead seaweed brought
up by recent tides, but is rarely continuous and often fades out
completely. Indeed on many shores no vegetation can be found at
all in this position, and it may then be assumed that the surface of
the ground is too frequently disturbed for plants to become established.
It is inclined to have a rather " scruffy " and untidy appearance, but
should not be neglected on this score, since some of the rare seaside
plants like sea-radish (*Raphanus maritimus*), sea-kale (*Crambe maritima*)
or the handsome oyster-plant (*Mertensia maritima*) sometimes appear
in this zone. This belt of " strand " vegetation, as we may call it,
possesses sufficient individuality to merit a short chapter of its own,
but it will be best if we do not include in it any description of plants
growing on rocks at this level, since these belong rather to the category
of cliff-vegetation, which is discussed in Chapter 9.

Perhaps the most characteristic communities are those found on
the dry sand at the top of beaches. They often recur in front of sand-
dunes (Pl. VII, p. 30), or sometimes at the base of cliffs normally
out of range of the high tides. A shingle beach, which is too unstable
to bear any vegetation, or which is largely submerged at high tide,
may frequently show a similar community of plants along its upper
limit, where the shingle is thinly spread over the soil of the mainland.

The shores of the innumerable sea-lochs and inlets along the western coasts of Scotland and Ireland provide long stretches of this type of habitat. In addition to these well-marked natural habitats, there are also certain artificial ones where the same kind of strand vegetation can sometimes be found. Large stretches of our coastline, particularly in the vicinity of harbours and coastal resorts, have been so completely altered by man that the natural vegetation has long since disappeared. Usually some sort of sea-wall has been constructed along the top of the beach, and the shore material often becomes piled up against this by the waves. It may thus be raised above the level reached by normal tides so that it becomes possible for plants to establish them- selves there. Exactly the same kind of plants are found in such places as in natural habitats, and in built-up areas these may provide the only trace of coastal vegetation to be seen for a long distance along the coast.

Although strand communities are found in a number of quite distinct habitats, they have certain factors in common. The most important of these is that they all occur at a position on the shore where considerable amounts of tidal drift are likely to be present. This drift consists of organic matter derived from decayed vegetation, mainly seaweed, which is deposited when the tide covers the shore. It is a valuable source of plant-food, and we shall see later (p. 125) that it plays a special part in maintaining the vegetation on sterile shingle beaches. The position of this belt of strand vegetation, when it occurs, is so constant that there can be little doubt that this supply of humus from the sea is largely responsible for maintaining plant-life in these inhospitable habitats. Plants growing on sand-dunes do not have the benefit of this source of plant-food, since they are out of reach of the tide.

Another characteristic of all these habitats is that they are likely to suffer periodically from considerable disturbance. For instance, in times of storm they may be swamped by sand or shingle, and are always liable to inundation by the highest spring tides. Foreshore communities are therefore decidedly ephemeral, and sometimes disappear altogether from a place where they seemed to be well- established. I remember having this forcibly brought home to me in Cornwall some years ago. I had undertaken to show a friend of mine a specimen of the sea-bindweed (*Calystegia soldanella*) (Pl. 6, p. 94) which he had never been able to find in flower. I knew the precise place at the top of a certain sandy beach where I had found it the previous year, and took him straight to it. But when we arrived, not only was there no sea-bindweed, but there was no foreshore vegetation to be seen at all—shifting sand had completely destroyed the whole community.

A third factor which is common to these habitats is that the material in which the plant has to grow always contains a certain amount of salt. This may not be very much, but it is never likely to be completely leached out by rain, since from time to time an exceptionally high tide will replenish the supply. Besides this, any plant growing on an exposed shore always has to contend with a certain amount of spray carried in by the wind. Most of the typical plants found are therefore halophytes, or at any rate have some tolerance of salt.

All foreshore communities are very open, with large areas of the available space not colonised at all. The mortality amongst seedlings must be extremely high, for when they have successfully germinated, a sudden change in the conditions may quite easily destroy the whole population. Nor does the composition of these open communities remain constant, but some species appear and others drop out as the conditions fluctuate. Most of the coastal plants which are found are annuals, and the characteristic species are not numerous. A certain number of common annual weeds, typical of inland waste places, are sometimes found with them. These plants are evidently able to tolerate the small amounts of salt in the ground during their season of growth, though an unusually high tide during the summer may well wipe them out completely. In addition to these, perennial halophytes may sometimes be found, but unless the habitat remains unusually stable, they rarely become permanently established. Taking the coastline as a whole, it can be said that a large number of the plants normally found in other coastal habitats *may* turn up casually in a foreshore community, although the number of regular inhabitants is always small. Alien plants, too, often appear in these communities, where the openness of the vegetation makes it easy for them to become temporarily established. This applies particularly to beaches near harbours, where, as a result of cargoes arriving from overseas, seeds of alien weeds are liable to be broadcast. I have often found rare plants in such places and can assure botanists who delight in tracking down unusual species that sea-shores, and indeed any waste ground near the sea, are well worth investigating.

As a rule, only three or four species will be found together along any particular stretch of shore, and often there may be only a single one which manages to keep a precarious hold at a few places, widely separated from each other. The largest number of different plants is likely to be seen when the top of the beach passes smoothly into the edge of the land behind it. Here there may be a thin zone composed of sand or shingle mixed up with some soil. With a more stable substratum and greatly improved water-holding capacity, denser vegetation, made up of a mixture of shore and inland plants, will probably develop.

Since casual plants are so frequently found, and also because of the general ephemeral nature of the vegetation, it is difficult to produce a satisfactory list of plants which are most typical of foreshores. Yet there are certainly some plants which seem to occur more frequently here than anywhere else. As far as sandy beaches are concerned, my own experience is that the prickly saltwort (*Salsola kali*) (Pl. I, p. 6) with its distinctive spiny leaves, and the sea-rocket (*Cakile maritima*) (Pl.XVI, p. 83) with its long straggly stems and pretty pink flowers, are by far the commonest inhabitants. In the West, at any rate, they often seem to be the only strand plants over long stretches of shore and are rarely found in any other position. Patches of certain species of orache (usually *Atriplex glabriuscula*) and odd plants of sea-beet (*Beta maritima*) (Pl. XXXVIII, p. 199) appear amongst them from time to time, but these may also be seen on shingle and rocks. Sometimes, too, plants more typical of open sand-dunes may join them, such as sea-spurge (*Euphorbia paralias*) (Pl. XIX, p. 102), sea-sandwort (*Honckenya peploides*), or sea-bindweed. On shingly beaches, the vegetation usually looks rather different, largely owing to the absence of prickly saltwort and sea-rocket. For long stretches of the shores of inlets on the west coasts of Scotland and Ireland, for instance, sea sandwort, orache species, sea-mayweed (*Matricaria maritima*) (Pl. XXXV, p. 178) and scurvy-grass (*Cochlearia* species) are the commonest seaside plants. These are usually mixed with certain inland plants, particularly silverweed (*Potentilla anserina*), goosegrass (*Galium aparine*), and chickweed (*Stellaria media*), which often occur in great abundance.

The following is a list of some purely coastal plants which seem to be generally typical of foreshores. Many of these occur equally frequently on sand and shingle, but those marked with an asterisk are more characteristic of sand:

sea-rocket*   ..   ..   ..   *Cakile maritima*
scurvy-grass species   ..   ..   *Cochlearia officinalis, C. danica, C. anglia*
                                        (Pl. XXXII, p. 167)

sea-radish   ..   ..   ..   *Raphanus maritimus*
sea-sandwort ..   ..   ..   *Honckenya peploides*
sea-campion   ..   ..   ..   *Silene maritima* (Pl. XXXIII, p. 174)
sea-wormwood   ..   ..   *Artemisia maritima* (Pl. XXXI, p. 166)
oyster plant or sea-lungwort ..   *Mertensia maritima*
sea-bindweed ..   ..   ..   *Calystegia soldanella*
sea-beet   ..   ..   ..   *Beta maritima*
orache species   ..   ..   *Atriplex glabriuscula, A. littoralis, A. patula,*
                                        A. hastata, etc.

prickly saltwort *   ..   ..   *Salsola kali*
knot-grass species *   ..   ..   *Polygonum raii, P. littorale,* etc.

The inland plants likely to be found are too varied to be worth listing in detail. Besides the species already referred to, the mouse-ear chickweeds (*Cerastium tetrandrum* and *C. semidecandrum*), bird's-foot trefoil (*Lotus corniculatus*), and the ubiquitous groundsel (*Senecio vulgaris*) seem to be particularly frequent.

The two grasses, sea couch-grass (*Agropyron junceiforme*) and sea lyme-grass (*Elymus arenarius*) might also have been included as typical plants in the above list. On beaches providing insufficient loose sand for these plants to consolidate into dunes, they sometimes occur in quantity, and may even produce a nearly continuous covering. Since, however, both these grasses often play a rather special part in building up low " foredunes " in front of the taller marram-grass dunes, they will be considered in the next chapter as part of the general sand-dune formation. Indeed, foreshore communities in general can be looked upon as the first stage in the normal succession of vegetation on sand (psammosere) leading to the development of sand-dunes. Many foreshore plants do, in fact collect miniature dunes round them when growing in loose sand, but their combined effect in stabilising the surface is so slight in comparison with that shown by either sea couch-grass or marram-grass, that it can hardly be treated seriously. Moreover, dunes develop quite normally when there is no strand community along their edge.

CHAPTER 7

SAND-DUNE VEGETATION

THERE IS no need to be a trained botanist to recognise sand-dune vegetation when one sees it, for the miniature mountain ranges with clumps of marram-grass growing over them are unmistakable. Indeed of all plant formations it is the most constant in general appearance. Marram-grass is a practically universal dominant species in the earlier phases of the colonisation of all areas of blown sand in this country, and this chiefly accounts for the strong family likeness between them all, wherever they occur. We shall see later, however, that there may be considerable differences between the ultimate vegetation which is developed on sand-dunes in different localities.

Our sand-dune areas vary greatly in size. Large stretches of dunes occur, for instance, along the Lancashire coast, the north Devon coast, along both sides of the Bristol Channel, and at Culbin, on the Moray Firth. None of these, however, compares in extent with those found along the French or Netherlands coasts. There are, in addition, many quite small areas of dunes along our coasts, which reproduce on a small scale exactly the same processes whereby the shifting sand is anchored by plants as can be seen taking place in the larger areas. Many little sandy bays, particularly in the West, possess such a strip of sand-dune vegetation at the top of the beach.

I always find sand-dunes particularly pleasant places in which to look for flowers. If they are well-developed, it is generally not difficult to find a warm spot, completely protected from the wind, on days when botanising on an open beach or a salt-marsh may be an unpleasant occupation. They also have the great advantage that, even after prolonged and heavy rain, the turf on the older dunes remains firm and agreeable to walk upon. Indeed it is largely as a result

Plate 6 Sea-bindweed, *Calystegia soldanella;* a plant of open dunes and shingle. Cornwall

of this that sand-dune areas make such excellent golf-courses, though they possess the added advantage that they usually provide an almost unlimited number of natural " hazards." Needless to say, botanists sometimes have cause to regret the inevitable modifications to the vegetation which result when " fairways " and " greens " are prepared !

For the ecologist, sand-dunes are especially interesting as being, like salt-marshes, some of the few places in this country where the whole series of stages in the colonisation of a bare habitat can be observed actively taking place at the same time. They are often equally inter-esting to the ordinary plant-hunter, for they are excellent places to search for rare plants, which have been preserved from destruction by agricultural operations. I can call to mind a number of occasions when I have unexpectedly come across a choice plant in the " rough " on a sea-side golf-course, when I was looking for my ball.

The number of different plants which may be found growing on sand-dunes is very large, and there are few times of the year when some plant cannot be found in flower. Most dune areas have at least one season of the year when they become a blaze of colour. Ragwort (*Senecio jacobaea*), for instance, is an almost universal inhabitant of sand-dunes, and when it is in bloom towards the end of July and in early August, the slopes of the hills become a mass of gold. Many other brightly coloured flowers may also grow in large patches; for example, when viper's bugloss (*Echium vulgare*) occurs, it sometimes produces a sheet of glorious blue, and if the yellow stonecrop (*Sedum acre*) is also present, its colour appears even brighter than that of the ragwort on account of its compact habit. On some dunes, particularly those in the North-East, you may be lucky enough to come across a good patch of the bloody cranesbill (*Geranium sanguineum*), which is a grand sight when in flower. Quite apart from these striking spashes of colour, the whole surface of the older dunes is usually gay for long periods with such common plants as yellow bedstraw (*Galium verum*), bird's-foot trefoil (*Lotus corniculatus*), and innumerable yellow " hawkweedy " composites. Yellow is certainly the prevailing colour in the latter part of the summer, and very gorgeous it can be on a fine evening when the sun is low.

There are various names which are given to sand-dune areas in different parts of the country. Thus on both sides of the Bristol Channel they are often called " Burrows," in Cornwall, " Towans," on the

Lancashire and Cheshire coast, and I think also elsewhere, " Meals " or " Meols ", and at Lowestoft, " Denes." The general name in Scotland is " Links," but in the Hebrides and the west Highlands " Machair " is a universal name for the fixed-dune pasture developed on the landward side of the open dunes.

There is one fundamental difference between the sand-dunes we find in this country and the great areas of blown sand which occur in deserts, and this is that the very existence of our dunes is due to plant-growth. Any obstacles, such as dead seaweed or flotsam washed up on a beach, will cause miniature dunes to be formed if a wind is carrying sand over them, but the height of the dune produced will be determined entirely by the size of the obstacle. It is only when a living plant is present, possessing the power of growing upwards as well as sideways as it becomes covered with sand, that a dune can continue to grow steadily in height. In many desert areas there is no permanent vegetation on the sand, and thus no such consolidation can take place. We have already seen (p. 6) that there is nothing inherently maritime about a sand-dune, but in the British Isles and at many places along the coast on the continent it is marram-grass which is largely responsible for the initial form taken by sand-dunes, though a number of plants possess similar, though less effective, powers of forming dunes. When we come to consider how the surface of the dunes is consolidated, a much larger number of plants is involved, but few of these would be able even to get a footing on the sand if it had not been first partially stabilised by the marram-grass.

The materials from which dunes are formed are usually brought up by on-shore winds from beaches which are exposed at low-tide. The top layers of sand on the shore soon become dry and incoherent when it is uncovered at low-water, and a strong wind blowing from the sea across it will collect a considerable amount of dry sand and carry it towards the land. This is a familiar sight if one walks along any well-drained beach in a strong wind at low tide. Even on that part of the shore which is covered by normal tides, the amount of sand which is moved is surprisingly large; the loose sand from the zone above the normal high-tide mark is, of course, shifted in much greater amounts. Indeed it can be distinctly unpleasant to walk along this part of a beach, facing a high wind, for one's face is exposed to a continuous bombardment by the sand particles which it carries. Even so, unless the beach is comparatively flat and extensive, the area of sand uncovered at low-water is usually insufficient for dunes of any size to be formed.

The nature of the dunes which are produced will clearly depend to a considerable extent on the direction of the prevailing winds. On the west coast, the normal westerly winds are on-shore, and the sand-dune ridges are usually formed at right angles to the direction from which they come. On the east coast, however, the common westerly winds are off-shore, and it is the less frequent easterly on-shore winds, particularly during gales, which are largely responsible for the growth of the dunes. Generally speaking, the most extensive dune areas in England are found along the west coast, but in Scotland they are found on the east side, particularly along the strip from the Moray Firth to Aberdeen.

In all fairly extensive areas of blown sand several ranges of dunes of different ages can generally be observed, usually running roughly parallel to each other. These are most evident in areas where dune-formation is still actively taking place along the seaward edge, as in the case of the Lancashire dunes. Here it is definitely known that during the last 300 years, a belt of dunes over a mile wide in some places has been established between Formby and the Ribble estuary, and the process is still going on. The valleys between the dune-ridges are sometimes produced when the wind blows strongly from an unusual quarter and carries away sand from the usually protected leeward side of the dunes, before their surface has had time to become consolidated by vegetation, but may originate in various other ways (see Chapter 2). These valleys or " slacks " vary greatly in width and extent, and most dune areas show a complicated tangle of highly irregular crests and depressions. It can frequently be noticed that the youngest ranges of actively growing mobile dunes are higher than the older ranges on their landward side, which have settled and become fairly stable. This gives a certain amount of protection to the rest of the area and facilitates the establishment of plants on the older dunes.

If for any reason the supply of sand to the front dunes is cut off, further growth naturally ceases. This state of affairs can be seen at a number of places along the coast and it will be noticed that the surface of the sand has become largely consolidated and fixed by vegetation right up to the tidal limit. This may sometimes be succeeded by a stage when the sea begins to gain on the land, and the seaward face of the dunes is eaten away by the highest tides. The photograph (Pl. XVII, p. 98) shows this happening along the face of some old well-established dunes along the east side of the Camel estuary on the north coast of Cornwall, when a sudden change in the main channel of the river some years ago caused rapid erosion of the base of the dunes. The vegetation here was sufficiently mature to contain a certain amount of scrub and an uprooted bush of blackthorn will be noticed half-way down the face of the little cliff. Incidentally, during the invasion scare in 1940, the edge of these dunes was ornamented with an elaborate system of barbed wire defences, erected by the local troops, but less than two years afterwards the whole area had been swallowed up by the sea and nothing remained of this work save an occasional tangle of wire on the beach. When erosion of this sort takes place, the sand will once more be thrown up on the foreshore by the waves, and from here will later be redistributed over the land. It will eventually be reconstituted in the form of dunes farther back, so that the net result will be a movement of the dunes towards the land. Where the coastline remains stationary over long periods, sand will tend to be added continuously to the same range of dunes, which may thus reach a great height. This state of affairs can best be seen along the Baltic coast, where in places dunes up to 200 feet in height are formed. Should the growth of the vegetation be insufficient to consolidate the surface of such dunes, they are liable to wander and to overwhelm the adjacent land. In the British Isles this is not the serious problem that it sometimes is on the continent, but even here there have been many cases in the past of settlements and valuable land being inundated by sand. The buried churches of St. Piran's (near Perranporth) and St. Enodoc, both in Cornwall, and the vanished city of Kenfig (Glamorgan) are well-known examples, whilst the famous wandering dunes at Culbin on the Moray Firth still give some ground for anxiety despite the elaborate measures which have been taken to consolidate them (Pl. VI, p. 23).

We must now consider a typical series of stages in the colonisation of completely bare sand which lead eventually to the establishment of the relatively stable vegetation to be found on the oldest dunes. It has already been mentioned (p. 30) that this natural succession of communities on sand is known collectively as the " psammosere."

Plate XVII Erosion of the seaward face of mature dunes to form a low sandy cliff. Camel estuary, Cornwall.

Ian Hepburn

a   *View in 1933.*

b   *View in 1943.*

Ian Hepburn

Plate XVIII   Growth of a sand-dune area, Camel estuary, Cornwall.

Foredunes

We saw in the last chapter that there is often a well-marked belt of strand vegetation along the drift-zone in front of the foremost dunes. It is, however, by no means a constant feature and the small effect it has in stabilising the sand is a purely local one. On the other hand, the sea couch-grass (*Agropyron junceiforme*) often plays an important part in preparing the way for the entry of the more vigorous marram-grass (*Ammophila arenaria*), though it is absent from a number of areas. It can stand up to a certain amount of immersion by sea-water, and is thus able to establish itself within the reach of spring tides. The sand here is intensely mobile, but the plant rapidly produces an extensive system of creeping rhizomes which bind the sand, and like marram-grass, it has the power of sending up new aerial shoots when it becomes covered by fresh sand. It is thus an admirable pioneer in this unstable habitat, but it does not possess the almost unlimited powers of upward growth which are shown by marram-grass, and spreads more readily sideways. As a result of this, the " foredunes " which it produces are never very high. Another dune-grass, the sea lyme-grass (*Elymus arenarius*) has much the same habit and may produce similar low foredunes, but it is more local in occurrence, and rarely colonises a long length of shore. When neither of these grasses is present, marram-grass is the first plant to colonise the mobile sand, but it is never found established so close to the high-tide mark. Some small areas of sand at the top of narrow beaches are entirely dominated by sea couch-grass, and in Ireland there are quite extensive dune areas where marram-grass does not occur. Elsewhere, as for instance along the Lancashire coast, well-developed foredunes of sea couch-grass occur along certain stretches whilst marram-grass is responsible for the front dunes along others.

The open *Agropyretum* of the foredunes does not as a rule contain many associated plants, since the sand between the tufts of grass is always extremely mobile and is fully exposed to violent winds. There are, in fact, sometimes no other plants at all to be seen over long stretches. When other plants are seen, they are usually species which are characteristic of sandy foreshores in general. Indeed, the zone of strand-plants often merges completely into the foredunes. The sea-sandwort (*Honckenya peploides*), which can itself form miniature dunes, is probably the most common associate. On the Somerset dunes, however, the most frequent plant occurring in the foredunes is the sand-sedge (*Carex arenaria*). Taking this consociation as a whole, the following plants seem to be most typical, though probably there is no area where they can all be found growing together:

yellow horned poppy	*Glaucium flavum* (Pl. IX, p. 54)
sea-rocket	*Cakile maritima* (Pl. XVI, p. 83)
sea-sandwort	*Honckenya (Arenaria) peploides*
orache species	*Atriplex glabriuscula, A. patula,* etc.
prickly saltwort	*Salsola kali* (Pl. I, p. 6)
sea-spurge	*Euphorbia paralias* (Pl. XIX, p. 102)
Portland spurge	*Euphorbia portlandica*
sand-sedge	*Carex arenaria* (Pl. X, p. 55)

In addition, casuals in the shape of common cornfield weeds may become temporarily established on the lee-side of foredunes.

OPEN AMMOPHILETUM

We now come to the main dunes which are produced almost entirely by the growth of marram-grass. If there are sea couch-grass foredunes in any area, the marram dunes can immediately be distinguished in the distance by their much greater height. Unless they are undergoing erosion by the sea, they will always be found well out of range of the tide and some portion of them, at any rate, will probably be in the process of active growth. As the sand accumulates on the seaward side, the marram-grass spreads through it, thus causing the dune to grow towards the sea. At the same time, the foredunes, if they are present, will gradually move in the same direction in front of the advancing marram dunes. When once marram-grass has become established in reasonably deep sand, its power of expansion in all directions is almost unlimited (see p. 51). The British dunes are comparatively humble in height compared with some that may be seen on the European mainland. Thus a height of about 60[1] feet would appear to be typical of the higher dunes in this country (the great dunes at Culbin are exceptional), whereas at many places along the Gascony coast of France they reach heights well over 100 feet, and

[1]Where considerably higher dunes occur, as at Merthyr Mawr in Glamorgan, the blown sand rests on a higher substratum of rock, etc.

even higher dunes are found along the Baltic coast. The speed with which marram-grass can spread is remarkable. The two photographs (Pl. XVIII, p. 99) of a small area along the Camel estuary in Cornwall show how much the grass can extend during a period of 10 years. This example can be taken as a typical and not exceptional rate of growth, for the supply of available sand from the beach and the shoals uncovered at low-tide was in no way unusual. The width of the new dune area produced during this time was from 80 to 100 yards, and at the same time the average level was raised about 25 feet. The remarkable powers possessed by marram-grass of fixing shifting sand make it an invaluable agent in preventing sand from spreading over the adjacent land in areas where dunes have a tendency to wander (Pl. VI, p. 23). Large-scale planting of this grass has been carried out for this purpose at a number of places along the coast on the Continent. Our dune areas in Britain are mostly too small for this ever to be a serious problem, but successful planting has been carried out at a number of points, notably on the Culbin sands and along the Lancashire coast. In the past a good deal of marram-grass was harvested each year for use as a thatching material for stacks and buildings, and also for making mats. In a number of places, however, it was found necessary to forbid this practice, since it prevented its proper upward growth and rendered the surface of the dunes dangerously unstable and liable to erosion.

While it is true to say that marram-grass is virtually the sole agent in producing and holding together these mobile dunes, it has very little effect on the actual surface of the sand. The steep seaward face of the first range of dunes is often almost completely bare, the clumps of grass confining themselves to the crests and the lee-side of the ridge (Pl. III, p. 14). The wind is thus free to remove sand from the range at will, except where the tufts of grass are established. If the supply of new sand which is blown off the shore is not at least equal in amount to that which is removed from the dune by the wind, the face of the dune will be eroded. When portions of these open marram dunes are blown away in this fashion, the whole complicated system of underground stems and roots is exposed to view, showing clearly how the dune was originally built up (Pl. XXI, p. 110). Such an excavation is called a " blow-out " and is a common sight in all dune areas. Even if wind-erosion does not produce a definite blow-out, it is obvious that a great deal of sand will be continually blown off the front dunes and deposited farther inland. If this falls within the main dune area it will become consolidated by the more advanced vegetation on the fixed dunes, but it can usually be noticed that the agricultural land adjoining an area of sand-dunes contains a high proportion of sand, even when the soil has not been developed on old dunes. The fixation of the surface of the sand is eventually achieved by other plants gradually establishing themselves between the clumps of marram-grass.

In the earliest stages, the colonisation of the constantly shifting sand is an extremely difficult business, and one rarely sees any plants on the seaward face of a dune which is actively growing. At its base there may be a few shore plants and if there are any foredunes there may, of course, be quite a number of plants on their lee-side, particularly if there is a pronounced trough to give a little shelter. Usually one must climb over the top of the growing range of dunes in order to see the first signs of colonisation. On the leeward slope, the surface of the sand is far less frequently disturbed by the wind, especially at its foot, and plants can become established with less risk of being buried in the shifting sand. The unstable nature of the surface is, however, by no means the only factor which makes this type of habitat a difficult one for plants to colonise. Since it is composed of nearly pure sand, it contains no appreciable amount of humus, which means that there will be a shortage of nitrates and some of the other salts required by most plants. This absence of humus and the coarseness of the sand particles themselves combine to render the surface very dry unless there are frequent showers. In addition to this, there is a wide variation in the temperature of the surface. In the day-time this often becomes so hot that it is difficult to bear the hand in it, but it cools rapidly at night, owing to its low water-content. All these factors combine to make the permanent establishment of seedlings a very chancy business. On the other hand, when once plants have survived the early stages and have developed sufficiently long roots, there is usually plenty of moisture in the lower layers of the sand. Where dune-valleys or slacks are well-developed, these often remain moist through-out the year, and in wet periods may collect extensive pools of open water. It will be noticed that many of the perennial dune-plants possess exceptionally long roots, which enable them to utilise these deep-seated water supplies. The origin of this water presents a problem of considerable interest; some of it may be derived from the internal

Plate XIX Early colonists of shifting sand: sea-holly, *Eryngium mari timum* (left), sea-spurge, *Euphorbia paralias* (centre), and marram-grass (right). Sandwich, Kent.

Plate XX   Natural scrub of sea-buckthorn, *Hippophae rhamnoides*, developed on mature sand-dunes.   Wells, Norfolk.

formation of dew, which almost certainly takes place in the case of shingle beaches. This process is discussed more fully in the next chapter (p. 124).

The first colonists in the open *Ammophiletum* are chiefly non-maritime species. They are usually common inland plants, whose seeds are dispersed by wind, and many of them are just as likely to turn up on any other waste ground in the neighbourhood. The plants which arrive vary considerably from place to place, depending on the accessibility of their seed-parents. They are in any case decidedly inconstant, since they possess no means by which they can deal with the sand if it buries them. In addition, if they do not get damp weather at the right time, many seedlings may be destroyed soon after they have germinated. It will be noticed that a number of these plants adopt a rosette habit, with a circle of leaves lying close to the surface of the sand. This helps to prevent sand from being blown away and also restricts the transpiration of the leaves. Others commonly occur in a prostrate form and often develop extensive underground stems. In addition to these common inland plants, a number of plants which are largely confined to sand-dunes will usually be found. These include the usual strand-plants of sandy shores, particularly the sand-sedge, which often spreads rapidly. The sand-fescue (*Festuca rubra* var: *arenaria*) is nearly always the first grass to join the *Ammophiletum*, though I have usually found that it does not appear with the earliest pioneers, but begins to spread only when the surface has already acquired a partial covering of vegetation. In some areas, however, it is said to be a genuine pioneer. Both these plants are extremely efficient sand-binders, since they develop elaborate creeping rhizomes. The following is a list of some of the most frequent pioneer species:

COASTAL PLANTS:

sea-sandwort ..	.. ..	*Honckenya peploides*
sea-holly ..	.. ..	*Eryngium maritimum* (Pl. 1, p. 35)
sea-bindweed ..	.. ..	*Calystegia soldanella* (Pl. 6, p. 94)
sea-spurge ..	.. ..	*Euphorbia paralias*
Portland spurge	.. ..	*Euphorbia portlandica*
sand-sedge ..	.. ..	*Carex arenaria*
sand-fescue ..	.. ..	*Festuca rubra* var: *arenaria*

INLAND PLANTS:

storksbill	.. .. ..	*Erodium cicutarium* (agg.)
yellow stonecrop	.. ..	*Sedum acre*
groundsel	.. .. ..	*Senecio vulgaris*
ragwort	.. .. ..	*Senecio jacobaea*
spear-thistle	.. .. ..	*Cirsium vulgare* (*lanceolatum*)
creeping thistle	.. ..	*Cirsium arvense*
carline thistle	.. ..	*Carlina vulgaris*
smooth hawksbeard	.. ..	*Crepis capillaris* (*virens*)
umbellate hawkweed	..	*Hieracium umbellatum* (agg.)
cat's-ear	.. .. ..	*Hypochaeris radicata*
hairy hawkbit	.. ..	*Leontodon leysseri* (*hirtus*)
red-fruited dandelion	.. ..	*Taraxacum laevigatum* (*erythrospermum*)
scarlet pimpernel	.. ..	*Anagallis arvensis* (Pl. 2a, p. 50)

PARTIALLY FIXED DUNES

As a result of the decay of the vegetative parts of these first colonists, a certain amount of humus is added to the surface-sand, resulting in increased fertility and water-holding power. The dead leaves of the marram-grass itself are also responsible for the addition of much organic matter. It thus becomes possible for a wider variety of plants to grow, and in particular, mosses start to appear. Mosses are the most efficient stabilisers of the surface-sand when they are once established but, owing to their very short rhizoids (or " roots ") they cannot easily grow on a dry and unstable surface. It is noticeable that they usually appear earlier in the succession on the west coast and in Ireland than they do on the east coast, doubtless as a result of the greater humidity of the air in the West. The most universal moss is *Tortula ruraliformis*, a species which is confined to sand-dunes. In many dune areas, *Camptothecium lutescens*, a markedly lime-loving species, is also widespread. Both these mosses possess xerophytic devices in the shape of folded leaves with rolled-back margins, to assist their survival during dry periods. Elsewhere, *Bracythecium albicans* is often found in place of *Camptothecium lutescens*. Other characteristic mosses in the early stages of colonisation are various species of *Bryum*, such as *Bryum argenteum*, *B. capillare* and *B. pendulum*.

In this later phase of the *Ammophiletum*, when the surface is beginning to become stable, a wide variety of plants can generally be found.

These include the so-called winter annuals referred to in Chapter 4, whose shallow roots occupy the upper layers of the sand, where the humus is mainly present, and also a number of more deep-rooted perennials. Needless to say, the pioneer colonists are mostly still prominent in the vegetation at this stage. The following is a list of some of the additional species which are likely to be found:

WINTER ANNUALS:

whitlow-grass	*Erophila verna* (agg.)
mouse-ear chickweeds	*Cerastium semidecandrum* and *C. tetrandrum*
sand-chickweed	*Stellaria apetala* (*boraeana*)
thyme-leaved sandwort	*Arenaria serpyllifolia*
rue-leaved saxifrage	*Saxifraga tridactylites*
field madder	*Sherardia arvensis*
lamb's lettuce or cornsalad	*Valerianella locusta* (*olitoria*)
early forget-me-not	*Myosotis hispida* (*collina*)
speedwell species	*Veronica arvensis* and *V. agrestis*
sand catstail	*Phleum arenarium*
hair-grasses	*Aira praecox* and *A. caryophyllea*

PERENNIALS:

rest-harrow	*Ononis repens*
white clover	*Trifolium repens*
bird's-foot trefoil	*Lotus corniculatus*
silverweed	*Potentilla anserina*
yellow bedstraw	*Galium verum*
mouse-ear hawkweed	*Hieracium pilosella*

FIXED DUNES

With the arrival of mosses and such plants as are listed above, the surface of the sand gradually becomes consolidated and largely covered by vegetation. Later on, when still further plants have appeared, a nearly continuous carpet of vegetation is produced and we arrive at what is called a " fixed dune." The gradual process of colonisation from bare sand to fixed dune can readily be appreciated if one walks slowly inland in a straight line at right-angles to the first ridge of open dunes. Though I have done this on a good many sand-dunes, I always find it a most fascinating experience whenever I explore a new area. It can be even more interesting if one makes an attempt, as one moves farther inland, to estimate the frequencies of the plants which occur in the zones of different ages. If, however, one relies solely on a general impression when doing this, one is often quite wrong. The easiest way to arrive at an answer is to make a list of all the species which are found in (say) fifty 1-foot squares, selected at random at the same distance from the shore. If the number of squares in which a particular plant is found is now doubled, a " frequency percentage " for that plant can be obtained. The combined results for all the plants which occur give a much more accurate picture of their frequencies than can be obtained by guesswork. This method is more useful in analysing the closed vegetation on fixed dunes, which is reasonably regular, than the open communities found on mobile dunes, which are very variable, but it always gives a more accurate result than can be obtained by mere inspection, and does not take so long to carry out as might be imagined.

In some dune areas, lichens play an important part in the fixation of the surface, though in others they seem to be almost completely absent. As a rule, they do not appear until the surface has reached a relatively advanced stage in colonisation, as they require a protected and largely stable surface for their successful growth. The pioneer mosses seem on the whole to be rather more tolerant and to appear somewhat earlier in the succession, though in some protected areas both lichens and mosses are said to be able to establish themselves successfully on relatively bare sand. The most usual lichens to appear are species of *Peltigera* and *Claydonia*, and when these occur thickly the sand-hills often acquire a pronounced grey colour when seen in the distance. For this reason, dunes whose vegetation is nearly continuous are often referred to as " grey " dunes, as distinct from the young relatively mobile ones, which are described as " yellow " or " white " dunes. The actual colour is dependent on the presence or absence of iron salts; if iron is present and there is not much calcium carbonate, its colour is generally yellow; if the sand contains a large amount of calcium carbonate in the form of shell fragments, it usually appears white.

As soon as the surface of the sand has become more or less completely covered with vegetation, one notices at once a change in the appearance of the marram-grass. It is no longer the luxuriant plant it is on the open dunes, but looks unhappy and often has a number of dead leaves round it. It generally manages to survive in the fixed dunes for some years in a more or less stationary state, but its tufts look moribund and rarely flower. The contrast between the virile rapidly expanding plants growing in the open sand and these dying relicts in the fixed sward cannot possibly be overlooked. The precise reason for this loss of vigour on the part of marram-grass is not known. There can be no question of the surrounding vegetation swamping it, for it is still the tallest plant present in the community. It may be that certain toxic products are formed by the roots of the new vegetation, which are inimical to the marram-grass. At any rate, there can be no doubt that it is a plant which can only flourish when abundant aeration of its roots is possible. In much the same way, the shrubby seablite (*Suaeda fruticosa*) can flourish happily only in mobile shingle (see p. 53). A number of the other pioneer colonists are choked out when the surface of the sand becomes covered, amongst which may be mentioned the sea-sandwort, the sea-spurge, the sea-bindweed and some of the winter annuals. The dominant species in these fixed dune communities vary considerably from place to place; in many places the sand-sedge or the sand-fescue replaces marram-grass as dominant. Usually there are several species which share the dominance and in the West, mosses often make up a large proportion of the surface-cover. It will, however, be noticed that the vegetation in extensive fixed dune areas is by no means uniform, and many local societies, dominated by a number of different plants, can usually be recognised. Nor is the whole surface always completely covered, and one often comes across small relatively bare patches.

The plants making up the fixed dune vegetation are numerous and extremely varied. Most of those already mentioned as characteristic of *Ammophiletum* can usually still be found, but many other inland plants of various categories will be found to have joined them. In particular, a number of new grasses generally appear, and many more mosses. Though the carpet of vegetation appears continuous, there is usually enough space for a number of casual weeds from nearby fields to become at any rate temporarily established, some of which may be found adopting a very prostrate habit, which gives them a surprisingly different appearance.

There is one characteristic of the vegetation produced in the early stages of colonisation of most dune areas which is of some interest. It will be noticed that it always contains a strong element of calcicolous (lime-loving) species. This is naturally more pronounced in sand containing large amounts of calcium carbonate, but is quite marked even when the amount is small. It has already been pointed out

(p. 28) that all fresh sea-sand has a slightly alkaline reaction as a result of the salt which it contains. It will continue to be alkaline when it is part of the surface of a dune as long as it is exposed to the deposition of more sand from the shore whenever the wind blows strongly. When, however, a new range of dunes grows up on the seaward side, it forms a barrier which largely cuts off the supply of fresh sand, and the salt is then soon washed out by the rain. Apart from an alkaline or neutral soil, the chief requirement of calcicolous plants in general seems to be an essentially dry habitat. Young sand-dunes therefore fulfil the principal conditions. The following is a list of some of the markedly calcicolous plants which occur on immature sand-dunes in various parts of the country:

purging flax	*Linum catharticum*
kidney vetch or lady's fingers	*Anthyllis vulneraria*
purple milk-vetch	*Astragalus danicus*
salad burnet	*Poterium sanguisorba*
blue fleabane	*Erigeron acris*
ploughman's spikenard ..	*Inula conyza*
carline thistle	*Carlina vulgaris*
yellow-wort	*Blackstonia perfoliata*
field gentian	*Gentianella campestris* (agg.)
autumn gentian	*Gentianella amarella* (agg.)
viper's bugloss	*Echium vulgare*
lady's tresses	*Spiranthes spiralis* (*autumnalis*)
pyramidal orchid	*Anacamptis pyramidalis*
crested hair-grass	*Koeleria gracilis* (*cristata*)
and certain mosses, such as ..	*Camptothecium lutescens*

In looking through the lists of plants which have been recorded from more or less fixed dunes in fifteen different localities, I found the names of over 250 different species, exclusive of mosses and lichens. Naturally the presence of many of these plants is quite fortuitous and depends on the type of vegetation existing in the immediate neighbour-hood of the dunes. Most of them are clearly plants which find the conditions in the dunes just as suitable to their requirements as those in their usual habitats, and probably encounter less active competition there. There are, however, a number of plants which seem to occur with great regularity on most dunes. The following list has been compiled from the lists of species found in ten different areas. All the plants appearing below were found in 90 per cent of the localities considered and most of them were present in all:

mouse-ear chickweeds	✓ ..	*Cerastium semidecandrum, C. tetrandrum* and *C. vulgatum*
storksbill ..	..	*Erodium cicutarium* (agg.)
bird's-foot trefoil	.. ✓ ..	*Lotus corniculatus*
rose species ..	.. ✓ ..	*Rosa* spp. (usually *Rosa spinosissima*)
yellow bedstraw	.. ✓ ..	*Galium verum*
ragwort ..	.. ✓ ..	*Senecio jacobaea*
spear-thistle ..	..	*Cirsium vulgare (lanceolatum)*
creeping thistle	..	*Cirsium arvense*
mouse-ear hawkweed ..	..	*Hieracium pilosella*
cat's-ear ..	..	*Hypochaeris radicata*
hairy hawkbit ..	..	*Leontodon leysseri (hirtus* or *nudicaulis)*
red-fruited dandelion ..	..	*Taraxacum laevigatum (erythrospermum)*
centaury ..	..	*Centaurium minus (umbellatum)*
early forget-me-not ..	✓ ..	*Myosotis hispida*
eyebright ..	.. ✓ ..	*Euphrasia officinalis* (agg.)
thyme ..	.. ✓ ..	*Thymus serpyllum* (agg.)
buck's-horn plantain ..	✓ ..	*Plantago coronopus*
curled dock ..	.. ✓ ..	*Rumex crispus*
sand-sedge ..	.. ✓ ..	*Carex arenaria*
sand-catstail ..	.. ✓ ..	*Phleum arenarium*
sand-fescue ..	.. ✓ ..	*Festuca rubra* var: *arenaria*
marram-grass ..	.. ✓ ..	*Ammophila arenaria*

It will be noticed that only five of the above species are specifically coastal plants, and some of these were probably holding their own only with some difficulty. We have, in fact, arrived at a type of vegetation composed almost entirely of inland species.

If you lie down on a mature sand-dune and examine the ground closely, you will generally find an enormous number of seedlings of many different plants competing with each other for the available space. In order to appreciate the richness and variety of the flora at first-hand, it is well worth while trying to make a full list of these. It does not take long to learn how to identify most plants when they are not in flower, and is rather fun to do. It may also prove of interest to see how many of the plants in the previous list can be found in your particular dune area. Most places have their local specialities, but this is no place to give a list of the many rare plants which may be found up and down the country. Mention, however, may be made of the local prevalence of the burnet rose (*Rosa spinosissima*) in the north-western and Welsh sand-dunes, the fetid iris (*Iris foetidissima*) in those along the Bristol Channel, and the sand-pansy (*Viola curtisii*) in Braunton Burrows (Devon) and the machair on some of the Hebrides.

I think it may be said that the stages in the colonisation of sand are roughly the same in all British sand-dune areas up to the point where the surface has acquired a fairly continuous covering of vegetation. The type of vegetation which develops subsequently, however, differs widely from place to place. It is governed chiefly by the nature of the sand and also by the effects of various biotic factors.

DUNE-HEATH

The composition of the sand round our shores varies greatly, particularly in the amount of calcium carbonate which it contains. Thus on the east coast at Blakeney (Norfolk) it contains less than 1 per cent, whereas in some places in the South-West there may be well over 50 per cent in the form of broken-up shells. Indeed, it has for long been a common practice in some parts of Cornwall to remove great quantities of sand from the beaches as a source of lime for some of the acid agricultural land nearby. Generally speaking, the sand on our western coasts contains more carbonate than that in the East, though the sand on the shore in front of the extensive Lancashire dunes contains only about 6 per cent. It is a well-known fact that the surface-layers of soil in chalk grassland often become somewhat acidic in the course of time. Precisely the same process takes place with sand containing only small amounts of calcium carbonate. This substance is practically insoluble in water, but it will dissolve in weak solutions of carbonic acid. Rain-water always contains some dissolved carbon dioxide, which it has acquired during its passage through the atmosphere. When it falls on the sand, it reacts with some of the calcium carbonate and transforms it into soluble calcium bicarbonate, which is then washed out of the surface-layers. We can represent this process by the following chemical equation:

$$CaCO_3 + CO_2 + H_2O = Ca(HCO_3)_2$$

In this way the carbonates in the upper layers of the sand are gradually removed, and, if the initial amount present was small, the top layer

Ian Hepburn

Plate XXI   A large "blow-out," showing recolonisation by marram-grass, and (left foreground) marram rhizomes exposed.   Kenfig Burrows, Glamorgan.

Ian Hepburn

Plate XXII A moist "slack" colonised by creeping willow, *Salix repens*, which has formed secondary

may soon lose most of its supply. As long as there is plenty of calcium carbonate in the sand, little permanent humus accumulates except quite near the surface, since an alkaline and well-aerated soil stimulates the rapid oxidation of plant-remains to carbon dioxide. As soon as the carbonates are removed, however, humus accumulates rapidly and the soil begins to become acid. This effect has been carefully studied in several areas in England where the sand had a low carbonate content. In each of these places it was shown that, although the sand on the youngest dunes contained practically no organic matter and was slightly alkaline, specimens from the older dunes showed a steadily increasing amount of humus and a steadily decreasing amount of calcium carbonate with their age, becoming at the same time progressively more acid.

When the sand becomes acid in this way, the normal succession leads to " dune-heath," a consociation which is dominated by either heather (*Calluna vulgaris*) or heath (*Erica cinerea*) or by both, and which does not differ greatly from typical inland heathland. A change in the vegetation on dunes which are becoming acidic is sometimes noticeable even before the plant cover has become quite continuous. The most obvious sign is a less varied flora, containing in particular fewer members of the *Papilioneae*, but including more plants which are typical of inland grass-heaths. The following list gives an idea of the sort of plants which may be found on dunes which are becoming acid, before heathland has been developed. A number of these may also be found on neutral or alkaline sand, but when they occur there, they are rarely so prominent:

dog violet	..	*Viola canina*
heath milkwort	..	*Polygala serpyllifolia*
tormentil	..	*Potentilla erecta (tormentilla)*
heath bedstraw	..	*Galium hercynicum (saxatile)*
sheep's-bit	..	*Jasione montana*
harebell	..	*Campanula rotundifolia*
wood sage	..	*Teucrium scorodonia*
sheep's sorrel	..	*Rumex acetosella*
field woodrush	..	*Luzula campestris*
common bent-grass	..	*Agrostis tenuis (vulgaris)*
early hair-grass	..	*Aira praecox*
Yorkshire fog	..	*Holcus lanatus*
sheep's fescue	..	*Festuca ovina*

The gradual succession to dune-heath begins with the appearance of an occasional plant of heather or heath, amongst the moribund marram-grass. As one moves farther inland towards the oldest dunes, the numbers of these steadily increase until one may reach an area where most of the surface is covered by them, with only scattered relicts of marram-grass remaining here and there and most of the other associates having a hard struggle for survival. It may be noticed that the heather and heath occur first only near the crests of the dune ridges. This is interesting, for they are naturally the places where the leaching of the carbonates by the rain will have been most effective. The same thing can sometimes be seen on chalk downs, heather beginning to grow on the summits, but not on the slopes of the hills. In most of the blown-sand areas where this succession occurs, the development of any large area of heathland has been prevented by various factors, chiefly by agricultural operations. Thus in the enormous dune area along the Lancashire coast, there are only a few small patches of heathland. The South Haven Peninsula (Dorset), the Ayreland of Bride (Isle of Man), Walney Island, near Barrow (Lancs.) and many of the sand-dune areas along the east coast of Scotland are some of the best-known places where the development of dune-heath can be observed. In some areas, considerable amounts of gorse (*Ulex europaeus*) (or in the West, *Ulex gallii*) may appear in the heathland. Elsewhere, it is sometimes invaded by bracken (*Pteridium aquilinum*) or by brambles (often *Rubus caesius*), both of which may become locally dominant. Typical associates which may be noticed in dune-heathland, in addition to many of the plants in the previous list, are the cross-leaved heath (*Erica tetralix*), the heath-rush (*Juncus squarrosus*) and such grasses as the wavy hair-grass (*Deschampsia flexuosa*) and mat-grass (*Nardus stricta*). In Scotland, the bilberries (*Vaccinium myrtillus* and *V. vitis-idaea*) are sometimes found. From the available evidence it would appear that about 300 years is the average time necessary for raw sand containing little calcium carbonate to develop into heathland.

The Culbin sands area is peculiar in that heather occurs as a dominant species on the flat firm sand which lies along the coastal side of the main dunes. Here it forms large circular mats up to a yard across, with occasional patches of heath-rush and other plants, the rest of the area remaining bare apart from mosses and lichens. The sand is so firm and stable on the shore that mosses (chiefly *Bryum* species) and lichens (chiefly *Peltigera* species) can establish themselves very early. The mobile dunes at Culbin are, however, quite unusual in that they occur more or less in the centre of the area, and consequently the succession is not always easy to follow. The normal development from open marram dunes here appears to be towards a partially closed community with sand-sedge as dominant, which then passes directly into dune-heath, in which heather and heath are co-dominants. Until recently this was the only area in the British Isles where the effects of wandering dunes on the neighbouring vegetation could be witnessed (Pl. VI, p. 23), but extensive afforestation is now being carried out there which should prevent any further movement of sand in the future.

Conifers have for long been employed on the Continent as sand-fixers on a large scale, particularly on the French coast. In the British Isles successful planting of old sand-dune areas has been carried out in a number of places, notably on the Southport dunes and on parts of the Norfolk coast. But the Culbin district is the only extensive area where the successful conquering of mobile sand by this method can be studied at the present time. It may be of interest to describe briefly how this is being carried out. As long ago as 1839 attempts were made to reclaim the area by afforestation, but many of the early plantations were buried by the shifting sand, whilst others were felled during the last two wars. The Forestry Commission began systematic planting in 1922, working from west to east, since the sand tends to be carried eastwards by the westerly winds. It is necessary, however, to stabilise the surface-sand to some extent or the young trees are liable to be buried. This can be achieved by planting marram-grass, sowing various grasses and weeds, or by "thatching." The latter process has been found the most successful, and consists in pegging down branches of birch, broom, heather, etc. which grow naturally in the neighbourhood, on the surface of the sand in the direction of the prevailing winds. Thinnings from the older plantations are also useful for this purpose. The young trees, 2-3 years old, are planted through this cover and usually establish themselves quickly. Corsican pine (*Pinus laricio*) does best on the open dunes, Scots pine (*Pinus sylvestris*) thrives on the flatter and more stable sand, and on shingle, whilst the shore pine (*Pinus contorta*) has proved the most suitable for planting on the low swampy areas. In other parts of the country, Austrian pine (*Pinus nigra*) has also been used with success in exposed positions. On the Culbins, the afforestation is proceeding so successfully that it seems likely that before many more years have elapsed, this interesting desert area will be a thing of the past. This may well be regretted by physiographers, but there will be the compensation of a valuable addition to the country's timber resources.

There can be little doubt that, if suitable seed-parents were available in the vicinity and there was no disturbance by agricultural operations, dune-heath would ultimately develop into forest. The normal succession on a well-drained acid soil would be to coniferous forest, as does in fact take place in the case of the semi-natural pinewoods on the southern heathlands in Surrey and Hampshire. There is no dune area in Britain where there is any evidence of this having taken place, although on the Culbin sands a scrub of self-sown birch, elder and willow (*Salix aurita*) has appeared in places.

DUNE-GRASSLAND

There is, however, no sign of heathland in many sand-dune areas and a more usual development produces what is called " dune-grassland." This is especially characteristic of the calcareous sand along our western coasts and in Ireland, and includes the extensive areas of " machair " in the Hebrides and west Highlands.

If the original sand contains a fairly large amount of calcium carbonate, the slow leaching action of the rain can make very little impression on it and the soil remains alkaline. As a result of this, the flora remains essentially calcicolous and the calcifuge (lime-hating) plants, characteristic of heathland, do not appear. It will also be noticed that many of these dunes carry an enormous population of snails (*mollusca*), and the ground is often thickly strewn with their empty shells. These must add quite a large amount of calcium carbonate to the surface and are thus another factor which counteracts the leaching action of the rain. The flora of these calcareous dunes is generally rich and varied, approximating to that found on ungrazed chalk or limestone grassland, which is so well described in J. E. Louseley's book in this series, but usually including a number of the earlier colonists which are typical of dry places in general. The chief difference between dune-grassland and ordinary fixed-dune vegetation is that the sward is usually much closer. This is mainly due to biotic factors, particularly the almost universal prevalence of rabbits. It will

Ian Hepburn

Plate XXIII Belt of shrubby sea-blite, *Suaeda fruticosa*, growing along the drift-line of the "Fleet" behind Chesil Beach, Dorset.

Plate XXIV An almost barren area of stable shingle with mats of prostrate broom, *Sarothamnus scoparius*

often be noticed that many of the plants have been nibbled, thus preventing their normal development and encouraging a low sward in which grasses predominate. Some dunes are used as sheep pastures, and in Scotland the machair is employed extensively for rearing cattle, a short turf being produced in both cases. Mowing also encourages a close vegetation, as can be well seen on the many seaside golf-courses round our coasts. Incidentally, the game of golf was first played on sand-dunes in Scotland and, as has already been mentioned, Scottish dune areas are frequently called "links." The marram-grass has long since been eliminated from the fairways of seaside golf-links, but old plants often survive in the "rough." It can be seen, therefore, that dune-grassland is maintained in a state of comparative stability in many places chiefly by artificial means.

DUNE-SCRUB

If the development of the taller plants is not restricted by grazing or mowing, dune-grassland may pass into some sort of scrub. In some areas quite a thick tangle of shrubs develops, and once more, if suitable seed-parents were available in the neighbourhood, there seems no reason why this should not ultimately pass into climax forest, although the constant exposure to violent winds might make this process difficult. The areas of dune-scrub in Britain are, however, never large and there is no locality where natural woodland has been produced. On the east coast, particularly in Norfolk and Kent, a well-marked zone of scrub, dominated by the sea-buckthorn (*Hippophae rhamnoides*) can be seen in places (Pl. XX, p. 103). This rather attractive shrub, with its distinctive silvery foliage and orange berries, is of only local occurrence as a native, but has frequently been planted as a sand-fixer in other parts of the country. On the Burnham and Berrow dunes in Somerset and at Merthyr Mawr in Glamorgan, for instance, it is now well-established (too well at Berrow!), but attempts to establish it on Braunton Burrows, Devon, have not proved successful. Its usual associates on the east coast are plants such as elder (*Sambucus nigra*), blackthorn (*Prunus spinosa*), honeysuckle (*Lonicera periclymenum*), or brambles (*Rubus* spp.), etc. Small areas of scrub are also found on west coast dunes, generally with blackthorn or hawthorn (*Crataegus monogyna*) as principal species. Privet (*Ligustrum vulgare*) is frequently common in the scrub, and there is often a tangle of brambles and wild

F.C.   I roses (especially the dwarf *Rosa spinosissima* in the North-West). It is an interesting fact that the majority of these scrub-plants carry berries and are thus bird-sown.

SECONDARY DUNE VEGETATION

Although the surface of fixed dune areas appears to be fairly stable, it is a common sight to find places where the surface has been broken through and bare sand has once more been exposed. Rabbits, for instance, can easily open up the thin layer of turf by burrowing, and thus enable the wind to get in and start a " blow-out," which may reach a considerable size (Pl. XXI, p. 110). When these natural excavations occur in old dune ridges, they make admirable bunkers on golf-links; indeed the artificial bunkers constructed on inland courses are an attempt to reproduce these natural features. In addition to the appearance of blow-outs, much sand is often blown on to fixed dune areas from the mobile dunes nearer the sea. Both these processes tend to produce new areas of bare sand, which in course of time usually become colonised once more. The recolonisation of these bare patches follows the same general succession that can be seen taking place on open dunes (Pl. III, p. 14). I have sometimes noticed old moribund plants of marram-grass springing once more into active life round the edge of a blow-out situated in the middle of a fixed dune area. Seedlings also arrive and spread over the bare sand, expanding and flowering normally until once more the surface-vegetation becomes too thick for their well-being. If the bare places are some distance from areas where marram-grass is in active growth, they are more usually recolonised by such plants as sand-sedge, sand-fescue or common bent-grass, all of which are very effective in anchoring the surface of loose sand. When the bare areas are not extensive, the secondary colonisation in some places may be initiated by mosses and lichens. In all cases, however, other species from the surrounding fixed dune vegetation soon follow when the surface has been sufficiently stabilised.

VEGETATION OF MOIST SLACKS

No account of sand-dunes would be complete without a reference to the very distinctive vegetation which is found in the dune-valleys or "slacks," particularly those on the west coast (Pl. XXII, p. 111). We have already referred to the fact that the sand in these valleys is usually moist, and that open water may actually be visible at certain times of the year. The most characteristic plant found in this habitat is the creeping willow (*Salix repens*), which is a plant normally found on damp heaths and moors. Although it usually establishes itself first in the moist troughs of the valleys, it often spreads rapidly from here over the surrounding dunes, forming a thick carpet of vegetation when exposed to sand which is being blown off the mobile dunes; it is perfectly capable of growing up through it, as it readily produces new roots to tap the water-supplies below the surface. The plant can therefore keep pace with the accumulation of sand in the same way as marram-grass, and sometimes forms secondary dunes as much as 15 feet in height. These *Salix* dunes are often a conspicuous feature in the wider slacks amongst the Lancashire sand-hills and elsewhere. It dominates a distinct consociation which can be called the *Salicetum repentis*. In the drier places the associated plants do not differ greatly from those found on ordinary mobile marram dunes, although sand-sedge and silverweed are often particularly abundant. In the older areas the decayed willow leaves furnish an unusually large amount of humus on the surface, which on the Lancashire dunes renders the habitat suitable for the establishment of wintergreen (*Pyrola rotundifolia*), which is quite common there. In addition, the scarce yellow bird's-nest (*Monotropa hypopithys*) can sometimes be found in similar spots.

In the valley-troughs, where there is most moisture, a community of damp-grassland and marsh species is usually associated with the dominant willow. This vegetation is sometimes referred to as " dune-marsh." Usually a large number of mosses, particularly species of *Hypnum*, is present. The only coastal plants which may be found here to give it a slightly maritime character are the sea-rush (*Juncus maritimus*) or the sea-milkwort (*Glaux maritima*). The following list gives an idea of the type of plant which may be found in these wet slacks:

cuckoo flower or lady's smock		*Cardamine pratensis*
grass of Parnassus	.. ..	*Parnassia palustris*
water-purslane	.. ..	*Peplis portula*
marsh pennywort*	.. ..	*Hydrocotyle vulgaris*
marsh bedstraw	.. ..	*Galium palustre*
yellow loosestrife	.. ..	*Lysimachia vulgaris*

bog pimpernel*	..	..	*Anagallis tenella*
brookweed	..	..	*Samolus valerandi*
marsh forget-me-not	..	..	*Myosotis caespitosa*
water-mint	..	..	*Mentha aquatica*
shoreweed	..	..	*Littorella uniflora*
marsh helleborine	..	..	*Epipactis palustrss*
marsh orchids	..	..	*Orchis latifolia, O. praetermissa, etc.*
yellow flag	..	..	*Iris pseudacorus*
rush species	..	..	*Juncus spp.*
marsh clubrush*	..	..	*Eleocharis palustris*
sedge species	..	..	*Carex spp.*
small club-moss	..	..	*Selaginella selaginoides*

Some of these plants, especially those marked with an asterisk, often form largely pure societies.

In order to give a composite picture of the great variety of the vegetation to be found in sand-dune areas, it has been necessary to generalise considerably. There is, of course, never any sharp line of demarcation between the various stages of colonisation, but each one passes gradually into the next. In addition, the vegetation is rarely uniform over a large area and numerous local societies can usually be recognised within the broad types of vegetation which have been described. It is, however, quite easy to trace the main course of the plant succession in any reasonably large area, if it is considered as a whole.

We can conveniently conclude this chapter by summarising the main stages in the colonisation of sand in the form of a diagram:

SUMMARY OF SAND-DUNE SUCCESSION (OR PSAMMOSERE)

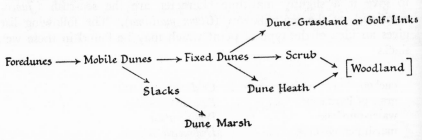

John Markham

Plate 7   Golden samphire, *Inula crithmoides*, on Portland Bill

CHAPTER 8

SHINGLE BEACH VEGETATION

S HINGLE beaches of one sort or another are a common feature all
round the coast, and it has been estimated that they make up at
least 300 miles of the shore-line of England and Wales alone. When
we talk about shingle, we mean any rock material which has been
thrown up on the shore by the waves and whose particles have to
some extent become rounded and smoothed in the process. The size
of the lumps of material of which it is composed varies from quite
large boulders to tiny pebbles, and the finest shingle approximates to
coarse sand. Generally speaking, however, shingle particles are of
sufficient size to roll freely over each other when they are set in motion
by the waves. Needless to say, whenever shingle is still exposed to
violent wave-action, the process of polishing and breaking down the
particles is actively taking place. The manner in which beach material
is sorted out and transported along the shore has already been discussed
in some detail in Chapter 2.

No one can pretend that these beaches are amongst the more
attractive portions of the coastline. They have a generally desolate
appearance, particularly when the inland scenery behind them is
featureless and uninteresting. Crabbe, who grew up on the Suffolk
coast, recalls the general grimness of the scene well when he writes:

> *Where all beside is pebbly length of shore,*
> *As far as eye can reach, it can discern no more.*

Walking along them for any distance is an unpleasant and exhausting
undertaking, particularly if you are in a hurry, and they are often
completely bare of vegetation for long stretches. Moreover they have

little to recommend them from the point of view of bathing; in rough weather they can be very dangerous, and at all times it is hateful to walk over the stones with bare feet before entering the water! When all this has been said, however, there remains much about them which is of interest to the field-botanists, and for those who like solitude, they are unrivalled. Not all shingle is bare, and in places where the surface is reasonably stable, some interesting and highly characteristic plants can usually be found. The unusual nature of the habitat makes such vegetation as there is peculiarly interesting and unlike that found anywhere else. At the right season of the year, too, if such plants as the yellow horned poppy (*Glaucium flavum*) (Pl. IX, p. 54), sea-campion (*Silene maritima*) (Pl., XXXIII p. 174), or yellow stonecrop (*Sedum acre*) are in flower and growing in large patches, the normally desolate shore springs temporarily into brilliant colour. The flora is by no means confined to coastal plants, for many inland plants establish themselves on shingle which has become stable, and in the absence of their usual competitors for space may spread rapidly and form large patches. I have, for instance, a vivid recollection of a certain Kentish beach which was one long stretch of glorious blue due to thousands of plants of viper's bugloss (*Echium vulgare*).

The methods by which shingle beaches come into existence have been sufficiently explained in Chapter 2. Here we need do no more than summarise the main types likely to be encountered. The simplest and also the commonest type in this country is the " fringing beach," which consists of a strip of shingle along the upper part of the shore, in contact with the mainland at all places. Examples of these are to be found all round the coastline, though the Channel coast exhibits some especially long stretches. The vegetation found on such beaches is generally confined to a comparatively narrow strip on the landward side of their crest. Far more interesting botanically are the " shingle spits " and " shingle bars." The former are usually produced at a point where there is a sudden change in the direction of the coast-line, and the shingle transported by the sea continues to be deposited along a line carrying on the original direction of the shore. This happens when the coastal currents pursue the same course and results in a peninsula growing out from the point on the shore where the current leaves it. The growing-end of such spits is frequently deflected land-wards in the form of a " hook." A succession of such hooks is sometimes produced by the continued growth of the spit in its original direction

(Fig. 13). When this occurs, it is clear that all but the last of these will be washed by relatively tranquil water, since the force of the waves will be broken by the hook farthest out from the shore. As a result, the shingle of which the older laterals is composed will become relatively stable, enabling a much wider range of plants to become established. Well-known examples of spits occur at Blakeney (Norfolk), Hurst Castle and Calshot (Hants), Northam (Devon), and Spurn Head (Yorkshire).

Shingle bars, on the other hand, are beaches running parallel to the shore-line, but at some distance away from it, and are usually joined to the mainland at each end. They sometimes enclose a back-water or lagoon. Bars of this kind may originate by the growing apex

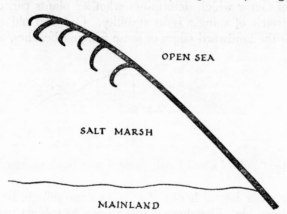

Fig. 13.—Diagram of a typical shingle spit, showing lateral hooks.

of a spit becoming attached to a projecting point of the mainland, or they may consist of the remains of a former land extension out to sea, which has been driven towards the land by wave-action. The outstanding example in England is Chesil Beach, which is over 10 miles long and encloses a long narrow lagoon, known as the "Fleet." Its width averages 500 feet, and it continues as a simple fringing beach for another 8 miles from its western end. Other good examples occur at Slapton (Devon), Pevensey (Sussex), Loe Bar (Cornwall) and Cemlyn (Anglesey). As far as vegetation is concerned, spits and bars present similar features, and most of the detailed ecological work on shingle vegetation has been carried out on these two types of beach.

Finally there is the " apposition beach," which is produced when shingle, instead of continuing its course along an existing spit or beach, piles up in front of it and is then driven by off-shore gales out of reach of the tides to form a bank parallel to one previously in contact with the sea. When this process is repeated a number of times, a series of roughly parallel ridges may be produced and an extensive area of completely stable shingle results. This formation is typically developed at Dungeness (Kent) and at Orford Ness (Suffolk). Apposition beaches differ from the preceding types in furnishing a stable habitat, and usually bear vegetation which is chiefly non-maritime. The number of species is considerably larger, and it may show the most advanced stage in the shingle succession.

The chief factor which determines whether plants can colonise a particular stretch of shingle is its stability. Only on old apposition banks, along the landward edges of some fringing beaches, and some-

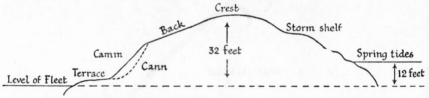

FIG. 14.—Profile of Chesil Beach (vertical scale much exaggerated).

times on the older lateral hooks of spits, can the shingle be described as completely stable. Elsewhere it will always be subject to periodical movement and, if the bank is a steep one, to almost continual disturbance. The degree of instability, however, varies greatly in different regions on the same beach, and it is chiefly for this reason that the distribution of the vegetation is usually so irregular.

It has already been pointed out that the movement of shingle may be brought about in a number of ways. Obviously the surface of the shingle on the seaward face of a bank, which comes under the direct influence of the waves, will always be kept on the move. During storms and exceptionally high tides, much shingle may be thrown on to the top of the bank or even right over the crest. Fig. 14 shows a typical profile for many shingle beaches, the flatter portion above the normal high-tide mark (known as the "storm-shelf") being the position where the bulk of the material is deposited on these occasions.

If the bank is not very high, it may become largely submerged during high tides, and much of the finer shingle may then run down the landward slope from the crest, spreading out in irregular tongues at its base to form a sort of terrace. It is clear, therefore, that direct wave-action tends to cause the whole bank to creep towards the land.

A similar movement landwards may also be produced by water percolating through a bank. This is a familiar property of many spits and bars, but is particularly well illustrated on Chesil Beach. It often happens that the level of the sea at high-tide is well above that of the water in the lagoon or salt-marsh channel on its landward side. If the bank is a high one, with steep sides, like Chesil Beach, the shingle may be considerably displaced by water passing through the bank from the higher to the lower level. This effect is unusually well-marked on Chesil Beach for, besides being exceptionally high, the water level of the Fleet varies very little owing to the narrowness of its outlet to the sea, and the difference in level on its two sides may reach 8 feet and over. Moreover, since the shingle composing the bank is particularly coarse and contains very few fine particles, the conditions for the passage of water through it are particularly favourable. Percolation, however, is never continuous through the whole of a bank, but is localised, with the result that a series of deeply-cut ravines (carns) are produced on the lee side, separated by conspicuous buttresses (camms). As a result of the passage of water through the ravines, a good deal of shingle is displaced from them and is shot out on the lee-side of the bank. These fan-shaped deposits may eventually unite to form the relatively level deposit known as the " terrace." The displacement of shingle naturally affects the backs and sides of the ravines, which are usually steep, the shingle rarely remaining stable for long but staying for short periods at its normal angle of rest. It is not surprising, therefore, that they remain completely bare of vegetation. The buttresses, on the other hand, have gentle slopes, and most of their surface remains fairly stable, with the result that quite a number of plants are usually found growing on them.

Shingle banks vary greatly in height and steepness, depending on the amount of available material and the height of the prevailing waves. Chesil Beach is exceptionally high, standing more than 30 feet above the level of the Fleet. This is no doubt due to the great force of the Atlantic breakers combined with the abundant supplies of shingle. The force of the waves at Westward Ho! (North Devon) is probably just as great, but the material available is scanty, and so the Northam Pebble Ridge is comparatively low. At Blakeney, on the other hand, there is an abundance of shingle but the height of the waves is considerably less than on the Atlantic coast, with the result that the height is only about 10 feet.

The consistency of the shingle not only varies greatly in different banks, but also in different parts of the same bank. On Chesil Beach it consists almost entirely of stones, but in other places it is often mixed with varying amounts of sand or mud. On most beaches, the storm-shelf consists of more or less pure shingle, and the smaller particles are likely to be found on the lateral extensions or the " terrace " on the lee-side (Fig. 14). Incidentally, the stones are automatically graded by the drift on Chesil Beach, the largest ones being found at the Portland end and the smallest near Abbotsbury.

One other small point should be mentioned. Some beaches undergo erosion on the landward side, if they impinge on tidal channels in their advance. When this happens, stable shingle cannot accumulate on the lee-side, and a certain amount of undercutting of their inner edges may occur. Wave-action may produce an analogous effect on exposed beaches round estuaries.

An unexpected characteristic of shingle beaches is the large amount of water they retain. Even in exceptionally dry summers the plants growing on them show no sign of suffering from drought. Obviously there will always be plenty of water in the lower zones on the seaward side which are washed by the sea each day, but this is salt water and any plants that grow near this part of a bank must be halophytes. It is not difficult to show, however, that there is abundant water at a higher level in the centre of the bank, and this water is fresh. Wells have been sunk in shingle at a number of places, and they often yield a copious supply of good water. You can prove the presence of water for yourself by removing the top layer of stones anywhere on the beach, when you will generally find that those only a few inches below the surface are quite damp. Moreover the presence of vegetation consisting largely of non-halophytes proves that this water is not markedly saline.

The origin of these abundant internal supplies of water presents a problem of great interest. We have already had occasion to refer to the same phenomenon in connection with sand-dunes (see p. 102) but it is much more remarkable that it should occur in such a coarse material as shingle. The coarseness of the medium naturally makes it easy for rain water to percolate deeply through it and accumulate in the lower layers. Moreover, the absence of small particles prevents this water from being subsequently drawn towards the surface by capillary action, so the loss from evaporation will be negligible. This, however, can hardly account for such abundant water, nor does it explain why the upper layers should remain damp. The most probable explanation of this phenomenon that has been put forward is that it is caused by the formation of dew within the bank. In the day-time, the surface of the shingle quickly gets warm in the sun and causes a certain amount of air to be expelled from the upper layers. This air is then replaced, partly by cooler air coming from over the adjoining sea or land, and partly by damp air drawn up from the interior. In the evening, when the heat of the sun is cut off, the stones cool down rapidly (their specific heat is low) and water is condensed on the surface of the pebbles from the damp air already surrounding them, and also from further supplies of moist air which are sucked in from outside. In this way moisture, partly derived from lower levels in the interior and partly from the outside air, arrives in the surface layers. If this is the correct explanation, it is probable that the same process takes place on a smaller scale in sand-dunes. Quite apart from this, in weather which favours condensation, appreciable quantities of water probably find their way into the interior from dew formed externally in the normal manner on a cool surface.

It is remarkable that the deep-seated water in shingle banks should remain fresh, for it almost certainly rests on a salt water-table. It has been shown, for instance, that the water-level in wells sunk in shingle definitely rises and falls with the advent of spring or neap tides. Evidently the diffusion of the salt upwards from the lower level of salt water must be so slow that the regular additions of moisture from above are sufficient to maintain a supply of fresh water on the surface. In Holland the similar accumulation of water beneath sand-dunes has been used on a large scale to supply the city of Amsterdam. Some care is necessary here to avoid pumping from too great a depth, lest the salt water at the lower level should be disturbed and cause the city's water supply to be contaminated with salt.

Despite the satisfactory nature of its water-supply, a shingle beach remains a sterile plant habitat if it contains no humus. Practically the only initial source of this is tidal drift, to which reference has already been made in Chapter 6 (see p. 90). The amount of this valuable material likely to be available to plants varies greatly with the type of beach, and also with the position on any particular beach. Thus when the shingle is completely cut off from the influence of the sea, as is the case on old apposition banks, there can be no further supply of drift. In these cases, unless humus can be supplied from other sources, there may be no vegetation at all. The only alternative source of supply is from the dead remains of lichens which sometimes become established on the surface of the stones. Shingle spits and bars, however, are much better placed, for they can receive their supply of humus from two directions. In the first place, there is the dead seaweed and other flotsam deposited by the waves on the seaward side, and later often scattered by the wind over the top of the crest. This source of supply is, of course, also available to simple fringing beaches. But there is in addition a supply of drift from the tidal channels or salt-marshes on the landward side, which may be equally important. This is not derived so much from seaweed as from the leaves and stems of plants growing on the mainland, the remains of the bodies of crabs and other animals, and even the dung of rabbits or farm stock which feed on the land vegetation. Although this drift is mainly deposited at the foot of the lee-slope, the higher spring tides will periodically raise it to a higher level. In course of time it becomes buried by the shingle falling down the slope above it and continues to decay steadily amongst the damp stones, causing the lower levels to become well permeated with humus. Since most beaches have a tendency to advance towards the land, it is clear that eventually this supply of humus can be spread throughout the width of the bank, though naturally the surface regions near the high-tide mark contain greater amounts than anywhere else. That there is a considerable amount of humus present in most shingle beaches is immediately apparent if you delve about in them, for you will soon find that your hands are quite dirty from the dark material on the surface of the stones. It is important to remember that the seeds of many of the plants which become established on shingle are carried up at the same time as the drift, and this is the principal method of dispersal in many cases.

The most detailed studies of shingle-vegetation have been carried out on the shingle spits at Blakeney Point and Scolt Head in Norfolk, both of which are now National Trust property. Each of these areas is particularly well situated to receive abundant supplies of drift from

John Markham

Plate 8 Tamarisk, *Tamarix gallica;* an "alien" shrub now familiar
near the coast

both sides. As a consequence of this, and also because of the proximity of sand-dune areas, their flora is distinctly more varied than that found on shingle elsewhere. Chesil Beach, Orford Ness, and indeed most other beaches anywhere round the coast have a much poorer flora and may be considered more typical.

The main topographical features of shingle spits and bars are shown in Fig. 14, p. 122, though there are few other examples where they can be so clearly distinguished as on Chesil Beach. The lowest portions of the seaward face of any beach, which often take the form of a series of steps, rarely show any vegetation at all. Even the rather flatter " storm shelf " higher up and well out of reach of all but quite exceptional tides, is usually bare, though it may support a few rather forlorn-looking foreshore plants like those described in Chapter 6, e.g. prostrate forms of orache or sea-beet. The " crest " of the bank is likewise generally bare, though a few stunted specimens of some of the plants found more happily ensconced in the " back " of the bank may have strayed there. It is obvious that the whole of the seaward slope of the bank is continually exposed to violent winds off the sea and to much salt spray, and so the establishment of young seedlings is bound to be difficult. The conditions on the " back " of the bank are quite different, since it is protected to a considerable extent from the prevailing winds and from much of the spray. The back generally takes the form of a gentle slope, unless there is much percolation of water through the bank, when, as we have seen, a series of steeply eroded ravines are produced. The gentle slopes and the buttresses between the ravines provide the chief region where the most typical shingle vegetation can be found. If there is a flat " terrace " along the base of the landward slope, a certain number of halophytes, which have escaped from adjacent salt-marsh areas, may sometimes be found mingling with the more characteristic shingle plants. When a number of lateral hooks, in the form of long fans of shingle stretching out towards the land, have been formed on a spit, a larger number of plants, many of which are not so characteristic of shingle, may eventually become established since their surface is relatively stable.

Shingle beach vegetation is always rather poor on species, as is only to be expected in such an inhospitable habitat, but there are a few plants which appear to be particularly characteristic. Only one of these, the shrubby seablite (*Suaeda fruticosa*), can be said to be more at home in shingle than anywhere else. It is a local plant, only abundant in a few localities in Dorset and Norfolk, and though it occasionally spreads into the adjoining salt-marshes, it can only flourish on the edges of creeks and in sandy places, where the drainage is good. It thus resembles marram-grass in requiring a well-aerated soil for its satisfactory growth, but, unlike it, finds shingle the best medium. Its characteristic habit when growing in mobile shingle has already been described on page 53. Its seeds are distributed by water and are deposited with the drift, with the result that if often occurs in well-marked littoral zones corresponding approximately to the level reached by the higher spring tides. Thus at Blakeney it forms a dense zone on both sides of the lateral " hooks," and also along the landward side of the main bank, except in those places where wave-action is too vigorous. Besides these, there are other less well defined zones higher up the bank, but these are less luxuriant, since they are much more exposed to the wind and there is less humus available. These higher zones may owe their existence to great storms in the past, which have thrown large quantities of stones right over the top of the crest and buried the zone of seablite growing along the lee-shore. Some of these partially buried plants may have later grown up through the layer of shingle covering them, while a fresh zone has been raised from seed brought up with the drift on the new shore-line lower down.

Chesil Beach is the other locality where the shrubby seablite is abundant, and here it forms a nearly continuous fringe on the terrace along most of the length of the Fleet (Pl. XXIII, p. 114). It shows no power of climbing up the steep slope of the bank and is practically confined to the drift-line. This is no doubt partly due to the much greater slope on the landward side of the bank and partly because the supply of humus is much less owing to the small movement of the tide within the Fleet.

All shingle beach communities are relatively open, and we cannot therefore speak of any species being dominant. Only on completely stable or dormant shingle, permanently out of reach of any tidal action, is a continuous cover of plants sometimes produced. The following is a list of some of the most characteristic plants found in open communities on shingle, taken from various parts of the country:

yellow horned poppy ..	..	*Glaucium flavum*
sea-kale	..	*Crambe maritima*
sea-campion	..	*Silene maritima*

sea-sandwort	..	..	*Honckenya (Arenaria) peploides*
purple cranesbill	..	..	*Geranium purpureum*
sea-pea ..	..	..	*Lathyrus maritimus*
yellow stonecrop	..	..	*Sedum acre*
oyster plant or sea-lungwort ..			*Mertensia maritima*
sea-bindweed ..	..	..	*Calystegia soldanella* (Pl. 6, p. 94)
woody nightshade or bitter-sweet (prostrate form)	..		*Solanum dulcamara* var: *marimun* (Pl. 2a, p. 50)
sea-beet	..	..	*Beta maritima* (Pl. XXXVIII, p. 199)
orache species	..	..	*Atriplex glabriuscula*, etc.
curled dock (maritime variety)			*Rumex crispus* var: *trigranulatus*
sea couch-grass species		..	*Agropyron pungens* and *A. junceiforme*

Some of the above plants are of decidedly local occurrence, but are included here because they are more frequently found growing on shingle in those few places where they occur.

Sea-campion is probably the most widely distributed shingle plant, often growing in large circular mats and making the whole surface gay in the spring with its dainty little white flowers. Its low habit of growth makes it able to withstand exposure to violent winds, and it appears to tolerate considerable amounts of spray. The sea-sandwort, although it is found most typically on open sandy foreshores, is also commonly found on any shingle containing some sand amongst the stones. Like the sea-campion, it forms large irregular mats of closely crowded shoots and its glossy succulent leaves do not appear to suffer from exposure to salt spray. Owing to their prostrate habit, both these plants may often be found spreading from the protection of the lee-side right up to the top of the crest. They develop unusually extensive rhizomes, which are extremely effective in binding the loose shingle, and both readily form new aerial shoots when buried by stones.

The maritime variety of the curled dock is another plant nearly always found on shingle beaches, and where it is well-established, produces a forest of tall, untidy-looking flowering stems in late summer. Unlike the two species just mentioned, its foliage is sensitive to salt spray; a sudden gale in the summer may destroy all the stem-leaves, though when this happens it usually shoots up again from the base later on. It is only vulnerable during the flowering season, when it has sent up a tall stem; at other times of the year it will usually be noticed that its base leaves are well protected by the dead remains of the previous season's growth. The handsome yellow poppy also produces an erect stem only during its flowering period, but its leaves are efficiently protected by their thick covering of hairs and it never appears to suffer from the effects of spray. It grows easily from seed but, although it behaves as a perennial when established on the less exposed parts of a beach, it rarely survives after its first flowering in an exposed position. Neither of the last two species possesses much power of holding mobile shingle, since they both grow from a deep vertical tap-root.

Yellow stonecrop is another plant peculiarly suited by its low growth to flourish on shingle. Although a common enough plant on dry ground inland, it is often found in large patches on shingle banks, usually fairly low down on the back, where it makes a brilliant display of gold when in bloom. Sea-bindweed is really more characteristic of open sand-dunes, but it may also be found on shingle in a number of places, particularly if some sand is present. Its prostrate habit enables it to survive even on the crest, well-established plants sometimes forming mats covering large areas. It is thus an effective stabiliser of the surface.

The pretty sea-pea, with its purplish-blue flowers, is a more local plant, nearly always found on shingle. It is very common on Chesil Beach, covering large areas of the back right up to the crest. It is also common at Aldeburgh and locally frequent on other Suffolk and Sussex beaches, generally preferring fairly mobile shingle. If cattle can stray on to the beach they devour it eagerly—a fact which may partially account for its limited distribution. Like other leguminous plants, it possesses the power of utilising atmospheric nitrogen directly to build up proteins, and no doubt its decayed remains provide a valuable source of combined nitrogen for other plants growing on the beach. The purple cranesbill is another local shingle plant, very common on the fairly stable shingle of the buttresses and the terrace on the landward side of Chesil Beach, and also on similar positions on some of the Sussex beaches. It closely resembles the common herb-robert (*Geranium robertianum*), also commonly seen on stable shingle, but it has smaller and rather more purple flowers and its leaves are markedly fleshy. The prostrate variety of the woody nightshade or bittersweet, with fleshy heart-shaped leaves, is also characteristic of south coast beaches.

The two maritime couch-grasses are more typical of the edges of

Plate XXV Broken chalk cliffs, isolated from direct influence of the sea, supporting chiefly inland vegetation. Branscombe Bay, South Devon.

Plate XXVI Flamborough Head from the air, showing the vegetation confined to the broken upper slopes of the cliffs

shingle areas. *Agropyron junceiforme*, as we have already seen, is primarily a pioneer on blown sand, but it often spreads into shingle if there are sand-dunes nearby. Sea lyme-grass (*Elymus arenarius*) sometimes behaves in a similar fashion. *Agropyron pungens*, on the other hand, is more characteristic of the upper edges of salt-marshes, but may likewise stray into shingle if it occurs on adjoining ground. The remaining species in our list, sea-beet, sea-kale and oyster-plant, are characteristic foreshore plants, and may sometimes be found on the seaward face not far above the high-tide mark. The two last are decidedly local, sea-kale being most frequent along the eastern half of the south coast, and the attractive oyster-plant or sea-lungwort being confined to certain beaches in Scotland and north-western England, apart from a small patch at Blakeney, which is its most southerly station in Europe.

The open vegetation of partially stabilised shingle shows no special pattern and varies greatly from place to place. On most beaches only a few of the plants given in our list will be found together and the whole vegetation may be made up of less than half a dozen species. In other places, particularly when salt-marshes or sand-dunes are present in the vicinity, many other plants may be found on the shingle. At Blakeney, for instance, numerous plants from the adjoining marshes and dunes have become established, and indeed that typical dune species, sea-holly (*Eryngium maritimum*), is chiefly found on the shingle there. Elsewhere, the sand-sedge (*Carex arenaria*) and even marram-grass (*Ammophila arenaria*), both typical of sand-dunes, are commonly found, whilst characteristic salt-marsh plants like sea-wormwood (*Artemisia maritima*) and sea-purslane (*Halimione portulacoides*) may be equally common in some districts. Besides these coastal plants, numerous inland species may also be seen, amongst which silverweed (*Potentilla anserina*), goosegrass (*Galium aparine*), the sow-thistles (*Sonchus oleraceus* and *S. arvensis*) and those universal weeds, chickweed (*Stellaria media*) and groundsel (*Senecio vulgaris*) may perhaps be singled out as being particularly frequent. In addition, if the surface is sufficiently stable and contains enough small particles, grasses such as red fescue (*Festuca rubra*), annual meadow-grass (*Poa annua*) or early hair-grass (*Aira praecox*) may establish themselves.

The later stages in the shingle succession can be seen only where the surface has become stable, as for instance on old apposition beaches and lateral hooks. Although a closed community of largely non-maritime vegetation may eventually be produced, there are remarkably few places where this can be observed. The intermediate stages, however, are sometimes well shown on the stabilised lateral banks or hooks on the lee-side of a shingle spit, as for example at Blakeney. We have already seen that these may often be well-supplied with humus and, moreover, may frequently consist of smaller stones than are typical of the main bank.

Although the vegetation on the Blakeney laterals is hardly typical, I think it is worth describing, as it occurs in well-marked zones and contains some interesting plants. The flanks of all the laterals slope gently, while the crest is flattened and only slightly convex. The lowest main zone consists of a dense mass of shrubby seablite, of a markedly different habit from that growing in mobile shingle, since in stable ground it remains upright. There is sometimes a zone below it of sea-wormwood and other salt-marsh plants, representing the upper edge of the marshland. Above the belt of seablite there is often an almost pure sward of red fescue. The most prominent plant in the next zone is the matted sea-lavender (*Limonium bellidifolium*), a smaller plant than the common species, with lilac-coloured flowers, which shows up as a vivid streak of mauve when in flower. Associated with it is a certain amount of the very local sea-heath (*Frankenia laevis*), and perhaps some thrift and sea-purslane. These three zones form comparatively narrow strips on the sloping sides of the banks. The uppermost zone is much wider and occupies the flat crest, which on the older hooks has generally accumulated a fair amount of soil. The most abundant plants here are thrift, sea-campion, and white bent-grass (*Agrostis stolonifera*), associated with a wide variety of subsidiary species which together make up a fairly continuous sward. Some of the principal associates are:

thyme-leaved sandwort	..	*Arenaria serpyllifolia*
mouse-ear chickweed ..	..	*Cerastium tetrandrum*
sea-pearlwort	..	*Sagina maritima*
yellow stonecrop	..	*Sedum acre*
buck's-horn plantain ..	..	*Plantago coronopus* (Pl. XXXVI, p. 179)
dwarf meadow-grass ..	..	*Desmazeria marina* (*loliacea*)
sea hard-grass	..	*Parapholis strigosa* (*Lepturus filiformis*)

The vegetation of the crest at the end of some of the oldest laterals contains an even larger number of species and is almost entirely non-maritime in character. Many of the plants mentioned in the last list are still present, but a number of others have joined them and some 50 different plants in all may be found. This is the most advanced stage in the succession to be seen at Blakeney. The only maritime plant found in any quantity is thrift, some of the other most abundant species being:

bird's-foot trefoil	..	..	Lotus corniculatus
yellow bedstraw	..	..	Galium verum
ribwort plantain	..	..	Plantago lanceolata
sheep's sorrel ..	..	..	Rumex acetosella
early hair-grass	..	..	Aira praecox
smooth-stalked meadow-grass			Poa pratensis
soft brome-grass	..	..	Bromus mollis

This zoning of vegetation is not usually so clearly marked as it is at Blakeney, nor can such a wide variety of plants be found, but it is universally true that whenever shingle ceases to be mobile (or is said to become " dormant "), a number of new plants usually appear. For instance, on the bank at Hurst Castle (Hants) and on certain east coast beaches golden samphire (*Inula crithmoides*) (Pl. 7, p. 119) forms a dense continuous zone along the shore-lines of the laterals in the same way as the shrubby seablite does at Blakeney, providing a lovely golden border when in flower. Viper's bugloss (*Echium vulgare*) is another attractive plant which often turns up on stable shingle, and in the absence of competition may spread widely. When a large mass is in bloom it is one of the most gorgeous sights to be seen anywhere along the coast. Crabbe refers to it pleasantly when he writes: " *There the blue bugloss paints the sterile soil.*" On a number of Scottish beaches I have seen scullcap (*Scutellaria galericulata*) making very attractive splashes of mauve, and the belt of sea-mayweed (*Matricaria maritima*), silverweed and other plants found on the stable shingle round so many Scottish sea-lochs has already been referred to in Chapter 6. The variety of plants which can become established is in fact endless, and whatever may be thought about the bleakness of mobile shingle, the vegetation on stable banks is rich in species and occasionally beautiful.

By the time this relatively advanced stage in the succession has

been reached, considerable amounts of soil are present on the surface
of the stones. Naturally when once a fairly continuous vegetation
cover has been established, humus will accumulate rapidly by the
decay of the vegetative parts of these plants when they have completed
their cycle. It must be remembered, however, that as soon as shingle
is removed from wave-action, its usual supply of humus from drift is
cut off, and it may become an even more sterile habitat unless there
is some alternative source of organic matter. There are, in fact, great
tracts of completely bare ground at Dungeness, Orford Ness and elsewhere, where there is insufficient humus for vegetation to become
established. Probably in many places where there is now plenty of
vegetation the initial supply of humus was furnished by lichens or
more rarely by mosses. Most lichens cannot grow on shingle that is
periodically mobile, and the presence of lichens on pebbles is usually
a sound criterion for the stability of a particular portion of a beach.
It is noticeable that lichens grow more luxuriantly on the south than
on the east coast, probably because the prevailing winds are both
milder and moister. Chesil Beach, for instance, carries a much larger
number of different lichens than Blakeney, chiefly on the terrace
above the Fleet, but the area of stable shingle here is also much larger.
Some lichens can endure exposure to sea-water better than others,
with the result that they often show well-marked zoning in relation
to different levels on the beach. Thus the black lichen, *Verrucaria
maura*, is generally found growing lowest, where it is subject to frequent
immersion by the higher tides; the attractive orange *Xanthoria parietina*
which is often such a prominent feature on sea-cliffs, is less tolerant
and is always found higher up. Mosses rarely appear until there is a
little soil, and are therefore not of comparable importance in making
the habitat fertile. When they occur, the species are much the same
as those found on stable sand-dunes and dry ground generally.

The most advanced stage reached by vegetation on shingle can be
seen best at Dungeness (Kent), where there is a vast accumulation of
material in the form of a number of apposition ridges covering an area
of about 10,000 acres. The area as a whole shows a great variety of
vegetation; large tracts are completely bare, whilst in other places a
patchy open community of shingle plants can be seen similar to that
found on the landward slopes of many normal shingle beaches. A
large part of the area, however, bears a relatively continuous type of
vegetation consisting chiefly of grasses and other inland species,

Plate 9 Sea-pinks, *Armeria maritima*, and sea-campion, *Silene maritima*, growing together on the edge of a Cornish cliff-top

maritime plants being represented by only occasional relicts. Much of this grass area is used for grazing sheep and has acquired quite a respectable amount of soil on the surface, though the original ridges or " fulls " are still clearly marked.

In a few places, where grazing has not interfered with it, a thin scrub has developed, composed of bushes of blackthorn, elder, gorse, hawthorn and brambles. Similar patches of scrub can be seen in several other shingle areas, notably on Calshot Point (Hants), suggesting that it might eventually be possible for the climatic climax vegetation (deciduous forest) to be developed, if suitable seed-parents were available in the vicinity and grazing were excluded. The almost continuous exposure to high winds, however, would always make the establishment of trees difficult, even if the scrub were sufficiently thick to afford some protection. The existing shrubs at Dungeness are all much distorted by the wind, the blackthorn and broom in particular often adopting an almost prostrate habit (Pl. XXIV, p. 115).

It may seem somewhat surprising that so much of this stable shingle should remain bare, but it must be borne in mind that the formation of soil in any quantity is a very slow process, particularly in such a well-aerated medium. We have already considered the effect on the fertility of the ground of cutting off the supply of drift from the sea, but isolation from tidal waters has another result. The seeds of a number of the most suitable potential colonists of bare shingle are water-borne, so that when the tide no longer reaches an area, only plants whose seeds are distributed by wind or birds (apart from lichens) have a chance of becoming established. There can be no doubt that those banks at Dungeness which have accumulated a fair amount of soil have been isolated from the sea for a very long time.

We have already seen that most shingle beaches, apart from those permanently isolated from the sea, have a tendency to move about, usually landwards, as a result of wave-action or the percolation of water through them at high tides. This movement can be quite rapid, and during severe storms large amounts of shingle may be shifted and the whole configuration of the beach completely altered. Where there is much shingle at the mouths of rivers, along the edges of reclaimed marshes or near coastal resorts there is always the danger that it may spread and cause serious damage. The shingling up of the port of Aldeburgh, which once had a flourishing shipbuilding industry, is a well-known example. Probably the most effective method of combating this danger is to establish a plant-cover to stabilise the surface, and with this end in view a certain amount of planting has been carried out from time to time at a number of points along the coast. Although a number of the common shingle-species possess distinct powers of binding the surface, the two plants found to be most efficient in arresting the movement of shingle are the shrubby seablite and tamarisk (*Tamarix gallica*) (Pl. 8, p. 126). The introduction of the former on the Northam pebble ridge, near Westward Ho! has been very successful in preventing further movement of shingle, which had previously been threatening the land behind it. Tamarisk, which is not a native of this country but is common in the Mediterranean region, has been planted with success in several places. It has, for instance, been effective in stabilising the crest of the shingle bank to the east of the harbour outlet at Shoreham (Sussex), and has been planted on the landward edge of the western end of Chesil Beach, beyond the termination of the Fleet, to restrain the spread of shingle over the adjacent farm-land. Tamarisk does well on sand too, and since it stands up bravely to continuous strong winds, is often employed as a hedge-plant in coastal districts in the South.

Shingle beaches are perhaps not so immediately attractive as most coastal habitats, for they are often bleak and their rather untidy appearance is against them, but they should not be neglected on this score. Those who are familiar with them in all weathers will know that they have an undefinable charm, which the sound of pebbles rolling back down a sloping beach with an outgoing wave will immediately conjure up for them. There still remains plenty of field-work to be carried out on their vegetation, particularly on that developed on stable shingle. In addition they provide yet another coastal habitat where it is possible to observe the successive stages whereby relatively advanced vegetation is produced from completely bare ground.

CHAPTER 9

VEGETATION OF SPRAY-WASHED
ROCKS AND CLIFFS

O F ALL maritime habitats, cliffs which are constantly exposed to salt spray and strong sea-winds are the most attractive and exciting. Here one sees nature at its wildest, and plant-life has its greatest struggle for survival. Wordsworth, describing the 'flowers growing over the entrance to Fingal's cave on the Isle of Staffa, writes of them:

> Hope smiled when your nativity was cast,
> Children of Summer! Ye fresh flowers that brave
> What Summer here escapes not, the fierce wave,
> And whole artillery of the western blast,
> Battering the Temple's front, its long-drawn nave
> Smiting, as if each moment were their last.
> But ye, bright flowers, on frieze and architrave
> Survive, and once again the Pile stands fast . . .

But cliff-vegetation has received remarkably little attention from ecologists, and there is an almost complete absence of cliff-studies in the literature. Perhaps the main reason for this is that for long stretches of the coast, the cliff-faces are completely inaccessible, and at other places they require a steady head and a taste for rock-climbing before any but distant observations can be made. Yet in many districts the line of cliffs does not remain continuously unbroken, but is periodically intersected by streams which have worn a way through to the sea. It is usually possible to descend to sea-level at their mouths, and by clambering about from these coves or beaches, one can often get a good idea of the vegetation on the surrounding cliffs. As this frequently remains much the same over considerable stretches, a satisfactory sample may be obtained in this way. Much can also be done with a pair of binoculars from a good vantage point if the cliff-face cannot be reached, since the vegetation is usually very sparse. Many of our cliffs, however, are either very broken or comparatively low, and these can be explored in detail more easily.

I blame myself greatly for neglecting this habitat in the past, when I had ample opportunity to study the cliffs in several different districts. I am by no means averse to a little mild rock-climbing if the rocks are sound, and would certainly have enjoyed botanising under such attractive conditions. As it was, I never realised until a few years ago how interesting a study of this vegetation might be, and then, rather belatedly, I did manage to carry out a fairly full preliminary study on a stretch of cliffs along the north Cornish coast. Since doing this, though I have visited a number of cliffs in other districts, I have never had long enough in one place to make more than a rough survey. A good deal of what is set out in this chapter is therefore based on my own Cornish observations, though I have managed to obtain some useful information about the cliffs in a number of widely separated districts through the kindness of various other observers. Any generalisations must of necessity be rather tentative, but I hope I shall succeed in giving an overall picture of this most attractive type of vegetation. It is scarcely necessary to point out that there is a great deal more work to be done on this habitat, and that any careful observations on cliff-vegetation, even if they are only lists of species, are likely to prove of value to future investigators.

The various ways in which the coastal cliffs of the British Isles originated have already been described in Chapter 2 (pp. 11-13), and it will be clear from what was said there that they differ widely in their degree of exposure to spray. Although long stretches of cliffs are washed by the sea at all tides, others may be situated quite a long distance back from the shore, usually as a result of alterations in the sea-level relative to that of the land in the past. Many of them are now separated from the sea by a well-marked platform or raised beach, while elsewhere the accumulation of beach-material at their foot prevents all but the highest tides from reaching them. It is obvious that the vegetation on cliffs set back some distance from the shore will only be indirectly influenced by the proximity of the sea, and it is not surprising that the majority of plants found on them are inland species

(Pl. XXV, p. 130). On the other hand, those which are directly exposed to spray bear characteristic communities of coastal plants, which conform to a fairly definite pattern. The variety of the vegetation found on cliffs which are isolated from most of the sea's influence is too great to be considered in detail here, and it must be clearly understood that this chapter is principally concerned with plants growing on cliffs which are fully exposed to sea-spray. It should, therefore, be pointed out that raised beaches, particularly when they are of rock, often provide a habitat similar to that found along the lower portions of exposed cliffs, and are usually excellent places for studying this type of vegetation.

Cliffs vary considerably in the amount of vegetation they can support. Those which are nearly perpendicular and are made of rock too hard to provide suitable crevices are sometimes completely bare to their tops. Others may be formed of such soft material that the surface is not sufficiently stable for plants to be permanently established, and may likewise be nearly bare. But on weathered and broken rocks, vegetation may often be found within a few feet of the water's edge, particularly if there is some protection from the prevailing winds.

Seen at its best, cliff-vegetation is to my mind the most beautiful of any that can be found round our coasts. Although the species of which it is composed differ considerably from place to place, I feel little doubt that its most constant and typical plant is thrift (*Armeria maritima*), or sea-pink to give it its prettier name. This is a beautiful species wherever it occurs, but when it is growing as a rock-plant and covers most of the surface of a cliff with its dainty pink heads in May, it is a glorious sight. Sometimes it is accompanied by sea-campion (*Silene maritima*), which flowers at the same time, and when thousands of its neat white blooms mingle with those of the thrift, it makes an enchanting picture (Pl. 9, p. 135). Later in the season, other cliff-plants may colour the rocks. I remember particularly a certain Cornish headland where great masses of rock sea-lavender (*Limonium binervosum*) (Pl. 15, p. 195) decked the foreground in misty blue, and in the distance the rocks were lit up by thousands of bright yellow stars of golden samphire (*Inula crithmoides*) (Pl. 7, p. 119) springing from its light green leaves. Scurvy-grass (*Cochlearia officinalis*), too (Pl. XXXII, p. 167), can be very beautiful when it covers the shadier ledges with dense masses of snowy-white flowers, and is especially acceptable as being always the first cliff-plant to come into flower.

Most of the plants found on the lower portions of cliffs are halo-phytes, which are also found in salt-marshes or on exposed beaches and foreshores. But there are just a few species which are largely confined to rocky cliffs in this country. Of these, samphire (*Crithmum maritimum*) (Pl. 10, p. 142) is by far the most abundant, though it thins out towards the North, and in Scotland is replaced by another umbellifer, Scottish lovage (*Ligusticum scoticum*), which, however, is never so widespread. Other purely cliff-plants are the wild cabbage (*Brassica oleracea*) (Pl. 13, p. 183), whose pale yellow flowers and large blueish-green leaves are a feature of some of the south coast cliffs, and the tall tree-mallow (*Lavatera arborea*), perhaps the most beautiful of all seaside plants, whose large pink flowers are pleasantly frequent along the Devon and Cornish coasts, but decidedly rare elsewhere. Nor must we forget the handsome sea-spleenwort (*Asplenium marinum*) (Pl. XL, p. 207), our only maritime fern, which can often be found in shady cracks along west coast cliffs. In addition, there are various species of sea-lavender, which are largely confined to maritime rocks, of which *Limonium binervosum* is the only one which is at all widespread, and one of the sea-spurreys, *Spergularia rupicola* (Pl. 11, p. 151), whose little pink flowers are a feature of many south-western and southern cliffs. Finally a mention must be made of that extremely vigorous "alien," generally known as "Mesembryanthemum" or Hottentot's fig (*Carpobrotus edulis*), which has escaped from gardens in the South-West, and established itself in dense masses on many cliffs. Its long trailing stems, with thick fleshy leaves, bearing rather garish magenta-coloured flowers, are quite unmistakable.

Many cliffs consist of a lower purely rocky portion, where the slope is fairly steep and there is little or no soil, and an upper, more broken, part, which is usually less steep and holds a certain amount of soil or detritus that has fallen from above. Unless they are quite low, the purely maritime vegetation is confined to the lower zones, the upper portions often carrying many inland plants belonging to the ground above, as well as various rock or shade-plants. When the cliffs are low, the whole vegetation may consist largely of maritime species, though it is nearly always thicker towards the top. Quite often there are no proper cliffs at all, but a belt of broken rocks instead, often in the form of a raised beach, forming a fringe to the grassland or moor-land, which is only a few feet above sea level. These spray-washed rocks, however, carry much the same sort of vegetation as that found on the lower portions of bigger cliffs, and are usually well supplied with crevices. Completely perpendicular cliffs without any ledges are rare, and when they occur there is little vegetation apart from lichens.

Besides differing widely in form and height, our cliffs show great variety in the materials of which they are made. Some rocks, like granite or basalt, are very hard, and unless much weathered may provide few suitable crevices and ledges for plants to occupy. Slaty cliffs, on the other hand, split easily, and those made of limestone usually possess plenty of ledges and are much eroded. If the material is a soft one, like clay or sand, the whole surface may be suitable for colonisation by plants, but the amount of vegetation will depend on its slope. If this is only slight, it may be thickly covered, but if it is at all steep it is likely to be too unstable to acquire a permanent plant-cover. There is also the chemical factor to be considered, for some materials are acidic and others are basic. As far as my own evidence goes, the chemical nature of the underlying rocks has little influence on the composition of the vegetation on the lower fully-exposed parts of cliffs, and it only becomes important in the upper zones, which do not carry purely maritime plant communities. The tops of limestone cliffs, for instance, usually show a more varied collection of plants than can be found on acidic rocks. But the amount of vegetation found along the lower zones depends in the first instance almost entirely on the physical properties of the cliff-material, since the ability of plants to grow at all will depend on there being sufficient soil. As far as rocks are concerned, the amount of available soil is directly related to the number of cracks and crevices that are present. Incidentally, this soil originates chiefly from detritus which has fallen from above, supplemented by the decayed remains of lichens and other plants (Pl. XXVI, p. 131).

But the absence of a sufficient number of crevices does not alone account for the general sparseness of the vegetation on the lower portions of all cliffs, and there are other factors which make it difficult for plants to become established. The most important of these is the continual exposure of this habitat to violent winds. It is often noticeable that the more protected a stretch of cliffs is from the wind (other conditions being equal), the thicker is the vegetation, and the lower it grows. Thus where the prevailing winds come from the west, east-facing cliffs show more vegetation, and there is usually a marked difference in the plant-life on the opposite sides of small islands. The primary effect of this violent exposure to wind is to dry out such moisture as is held by the small amounts of soil in the crevices. This in itself must make the establishment of fresh plants difficult, since they are likely to suffer in their early stages from an acute shortage of water, before they have developed sufficiently long roots to tap deeper sources of supply. In dry summers I have seen even well-established thrift plants withered up, despite their long roots. Most of the plants which are found in the lowest zone are perennials, eventually developing extensive woody roots which are also instrumental in anchoring them firmly against possible uprooting during gales. It is remarkable that the majority of the plants one finds seem to be old and long-established, young plants being relatively rare. This confirms the supposition that it is very hard for new seedlings to become established. The continual strong winds also encourage excessive transpiration from the leaves, and so restrict the normal development of the plants. This is the reason for the rather untidy and lop-sided appearance of much of the vegetation.

The strong winds are indirectly responsible for the other most important factor—the salt spray to which the whole cliff-face is to a greater or lesser extent exposed. At present we have little information about the amount of salt which can be tolerated by different plants. It is clear, however, that there are only relatively few species able to survive the combination of heavy spraying and exposure to wind characteristic of the lower portions of cliffs. All these must be able to tolerate high concentrations of salt on their leaves, for on a dry day after a storm one can often observe salt crystals on them. Naturally the true salt-marsh halophytes have no objection to this, but it would appear that a number of " spray halophytes " and even a few inland plants are also sufficiently tolerant. It would be interesting to make some measurements of the actual amounts of salt deposited during the year at various heights above the sea. With suitable permanent collecting vessels situated at different heights on a cliff-face, this ought not to present any insuperable difficulties.[1] Another possible line of approach would be to grow a number of cliff-plants in a garden plot, and to observe the effect of spraying them with varying amounts of salt water of different strengths over a considerable period.

[1]Some interesting experiments on these lines are recorded in a recent paper by E. T. Robertson (1951), *Trans. Bot. Soc. Edinb.* 35: 370.

F. Ballard

Plate 10   Samphire, *Crithmum maritimum*, on a red sandstone cliff, S. Devon

There will also, of course, be a certain amount of salt in the root-water, but this is even more difficult to estimate. The problem of obtaining a representative sample of soil from the interior of a narrow crack in some rocks is in itself considerable, and when a plant possesses roots several feet long, it requires a rather arbitrary decision to select a suitable depth for taking the sample! Moreover the variations in the salt-concentration near the surface must be very wide. Thus, when a big sea is running, the concentration may approximate to that of sea-water. If a period of sun and light breezes should follow, it will rise, as it does in the middle regions of salt-marshes which are only flooded periodically. On the other hand, if heavy rain should occur, it will fall correspondingly. It is unlikely that these variations of concentration in the surface-layers have an effect on well-established perennials, since their long roots enable them to derive their principal water-supplies from regions where the salinity is low and varies little. On the other hand it may be very important in controlling the establishment of young seedlings, which are not able to tolerate such high and variable amounts of salt. I carried out a few tentative estimations of the salt-content in the surface soil where the lowest vegetation occurred on my Cornish cliffs. Though these showed a rather wide variation between different specimens taken under the same conditions, they did, at any rate, confirm the supposition that the salt-content fell after rain, and rose in dry weather. Thus, on a dry day, when a high sea was running, the soil-water was found to contain 2.0-2.5 per cent of salt, but rose to a value of 2.5-3.5 per cent after three days of hot dry weather. On another occasion, after 24 hours of continuous heavy rain, the concentration had fallen to less than 0.5 per cent. Specimens taken higher up the cliffs showed generally lower results, but the same sort of variation with the weather. As my samples were all taken in the most exposed places, they probably show higher, rather than lower, values than the average, and it would be interesting to have some results from other cliffs.

It is difficult to assess the precise importance of the salt-factor, but I think it is clear that, as far as the lowest zone of vegetation is concerned, it is largely instrumental in determining which species can become established. Higher up, though its influence becomes steadily less, it still leaves its mark on the vegetation by excluding a good many plants. That this must be so is proved by the fact that some of the more tolerant inland plants, when growing on cliffs, acquire distinctly fleshy leaves as a result of absorbing salt (see p. 48).

If we examine a typical rocky cliff, whose base is permanently submerged in the sea, the lowest vegetation seen will be a purely marine zone of seaweeds, which are almost completely covered at high-tide. Not far above the high-water mark and well within the direct spray zone, the rocks are often thickly coated with certain maritime lichens, such as *Verrucaria maura* and *Lichina confinis*, which appear in the distance as a prominent dark band. As a rule these do not persist far from the water's edge, and higher up their place is soon taken by others. The common, but beautiful, *Xanthoria parietina* is the most widespread, often covering large areas of the higher rocks and giving them a lovely orange flush (sometimes assisted by *Placodium lobulatum*). Many other lichens are commonly found at all levels on a cliff-face, particularly noticeable being the white blotches of *Ochroleuca* (*Lecanora*) *parella*, and the large fluffy tufts of *Ramalina* species. Sometimes, chiefly on the west coast, one of the few maritime mosses, *Grimmia maritima*, may be found not far above the high-water mark, growing in rigid yellowish-green cushions. *Ulota phyllantha* and *Trichostomum mutabile* are also occasionally found close to the water.

The level where the first flowering plants appear varies greatly and depends on the sum of the factors which we have already discussed. On the north Cornwall cliffs, which are formed of slate and not exceptionally steep, there are usually plenty of suitable crevices and cracks, but I found that the lowest-growing plants occurred on the average at from 20 to 30 feet above the high-tide mark, when fully exposed to the prevailing wind. In sheltered coves they grow much lower, and this also applies to cliffs elsewhere not exposed to the full fury of the Atlantic. Needless to say, when the sea does not cover the base of the cliffs at all tides, or when there are rocks extending outwards horizontally to form a barrier from the sea, the vegetation begins lower still. But, at whatever level it occurs, this lowest vegetation is always rather sparse, and the number of different species found is small. It is remarkable, too, that some plants which are commonly seen in this zone are sometimes completely absent over long stretches only to reappear in equal abundance farther along. For instance in Cornwall, samphire is perhaps the commonest cliff-plant, but it is quite possible to explore half a mile of cliffs without finding it at all. In such open communities as these, it is impossible to speak of a dominant species, and competition for space does not arise. Indeed the whole vegetation often gives the impression of only just holding its own against the elements. Nor can one easily decide on the most abundant species, since this may easily alter from one stretch of cliffs to another.

After carefully examining some fifteen different areas along a fairly long stretch of the north Cornish coast, I found that there were only thirteen different plants ever to be seen in this lowest zone. From my more casual observations of cliffs in various other places, I fancy that this is a wider variety than is usually found in this position elsewhere, though apart from certain local rarities, nearly all the plants I saw there were also in my Cornish list. Since these seem to be reasonably typical, I give my thirteen species below, arranged in their approximate order of frequency:

thrift	*Armeria maritima*
samphire	*Crithmum maritimum*
rock sea-spurrey	*Spergularia rupicola*
sea-beet	*Beta maritima* (Pl. XXXVIII, p. 199)
orache	*Atriplex glabriuscula*
golden samphire	*Inula crithmoides*
sea-spleenwort	*Asplenium marinum*
buck's horn plantain	*Plantago coronopus* (Pl. XXXVI, p. 179)
rock sea-lavender	*Limonium binervosum*
scurvy-grass	*Cochlearia officinalis*
sea-plantain	*Plantago maritima* (Pl. XXXVII, p. 198)
sea-mayweed	*Matricaria maritima* (Pl. XXXV, p. 178)
sea-campion	*Silene maritima* (Pl. XXXIII, p. 174)

I do not feel in a position to give a list of all the other species found elsewhere in similar positions on cliffs, though the total number cannot be large. Most of them are decidedly local and some very rare. The following is a selection:

wild cabbage	*Brassica oleracea*
sea-kale	*Crambe maritima*
sea-radish	*Raphanus maritimus*
stalked scurvy-grass	*Cochlearia danica*
tree-mallow	*Lavatera arborea*
Scottish lovage	*Ligusticum scoticum*
sea-wormwood	*Artemisia maritima* (Pl. XXXI, p. 166)
sea-milkwort	*Glaux maritima*

In addition, grasses sometimes grow thickly on the lower ledges, usually *Festuca rubra*, or sometimes, in the North, *Puccinellia maritima*.

The vegetation higher up the cliffs, as has already been pointed out, usually shows a larger number of species and varies much more from place to place. Yet despite the fact that the surface is often more broken and there is less exposure to spray, it is rare to find anything approaching a closed community except quite locally. As a rule, most of the maritime plants found lower down are still found here, but they have been joined by a number of inland or submaritime species from the grassland or heathland on the cliff-tops. Besides these, inland " fissure " or wall-plants, such as pennywort (*Umbilicus rupestris*), (Pl. 11, p. 151) wall-pellitory (*Parietaria diffusa* (*ramiflora*)) or ivy (*Hedera helix*) are often found, the latter frequently spreading widely over the surface of the upper rocks. Shade or woodland species, too, such as angelica (*Angelica sylvestris*), honeysuckle (*Lonicera periclymenum*), hemp agrimony (*Eupatorium cannabinum*) and many others, may be found in the damper clefts. In Cornwall I found that all the "regulars " from the lower zone also occurred higher up. The most constant newcomers were sea-carrot (*Daucus gingidium*), (Pl. XXIX, p. 162) English stonecrop (*Sedum anglicum*) (Pl. 14, p. 190), kidney-vetch (*Anthyllis vulneraria*), and various grasses from the cliff-tops. In addition, sea-campion, which grew only sparsely on the lower rocks, became abundant. I have seen all these in similar positions on cliffs elsewhere, and I think they may be considered typical. The number of other species that may appear here is so large that I shall not attempt more than to mention the following small selection, which have struck me as being particularly frequent:

mouse-ear chickweed ..	..	*Cerastium tetrandrum*
sea-pearlwort ..	..	*Sagina maritima*
bird's-foot trefoil	..	*Lotus corniculatus*
blackthorn	..	*Prunus spinosa*
silverweed	..	*Potentilla anserina*
yellow stonecrop	..	*Sedum acre*
coltsfoot	..	*Tussilago farfara*
thyme	..	*Thymus serpyllum*
curled dock	..	*Rumex crispus*

Generally speaking, the more broken the cliffs and the less steep their slope, the greater the number of different species, and the thicker

John Markham

Plate XXVII A cliff-wall covered with sea-pinks, *Armeria maritima*. Cornwall.

John Markham

Plate XXVIII Steep cliff-top thickly carpeted with a mixture of maritime and inland plants: bluebells, red campion, thrift, sea-campion, etc. North Cornwall.

the general vegetation. Sometimes this passes gradually into that found on the cliff-tops, but when the cliff-face remains steep, a sharp break occurs between the open communities of the rocks and the relatively close sward on the top. In either case, as soon as there is a sufficient quantity of soil lying over the surface of the rock, the maritime species find themselves unable to face competition from the thickly growing inland plants which now appear, and they rarely survive far from the rocky parts of the cliff. The belt of mixed vegetation characteristic of the cliff-edges is discussed more fully in the next chapter.

Proof that this abrupt cessation of cliff-plants is due primarily to competition from stronger-growing inland species is sometimes provided by an inspection of the vegetation on nearby stone walls. In the West particularly, these have often been erected to divide the uncultivated ground of the cliff-tops from the fields on the landward side. They are usually roughly constructed of large boulders, sometimes simply piled on each other, but more often held together with a certain amount of earth. Though they may be situated some little distance from the edge of the cliffs, they nevertheless provide a somewhat similar habitat to that of a cliff-face. They do not hold sufficient soil for inland cliff-top plants to become at all thickly established, but they contain plenty of cracks and crevices for rock-plants. Moreover, they are highly exposed to wind, particularly if the ground slopes away towards the edge, as it often does. They must also receive appreciable amounts of spray in rough weather, for I have often noticed that windows of houses on high ground as much as a mile from the shore become coated with salt after a gale. The conditions are thus very suitable for the growth of cliff-plants, and it is not surprising to find that these walls sometimes carry nearly pure cliff-vegetation (Pl. XXVII, p. 146) and (Pl. 11, p. 151). I have seen walls as much as 500 yards from the edge, thickly colonised by thrift, sea-campion, scurvy-grass, sea-beet, sea-spurrey, etc. Oddly enough, I have never noticed samphire, despite the fact that it was often abundant on the cliffs below, but this may not be typical behaviour elsewhere. Those plants which become established seem to appreciate the slightly less severe conditions provided by this artificial habitat, and some of the finest displays of sea-pinks I have seen anywhere in Cornwall have been on these cliff-walls. Incidentally, those fully exposed to the prevailing winds always seem to be more thickly covered than those which are more protected. It would be interesting to have some observations from other districts, my own being confined to various parts of the West Country. The sea-walls which have been built along parts of the coast where there are no cliffs are also quite worth inspection. Provided they are not exposed to excessive amounts of spray and are sufficiently roughly put together to provide adequate cracks, they may support quite a number of cliff-plants.

In some places, the rocky ledges on precipitous cliffs are used as nesting places for enormous numbers of sea-birds. This is particularly true of certain small islands and inaccessible "stacks." The amount of excreta produced by a large colony of birds is very great, and it may modify the cliff-vegetation considerably, particularly at the base of the cliffs, if this should be isolated from the sea. Some cliff-plants become very luxuriant and grow exceptionally large under these conditions, notably scurvy-grass and sea-beet. I have also noticed unusually coarse-looking specimens of rock sea-spurrey, orache, and sea-mayweed in such places. In addition, a number of inland weeds not found elsewhere on the cliffs often turn up, particularly " nitrophilous " or nitrogen-loving plants. The following inland species have been noticed as being characteristic of these rocks:

chickweed	.. .. ..	*Stellaria media*
white campion ..	.. ..	*Melandrium album*
goosegrass	.. .. ..	*Galium aparine*
sorrel species ..	.. ..	*Rumex acetosa* and *R. acetosella*
nettle ..	.. .. ..	*Urtica dioica*
annual meadow-grass	..	*Poa annua*

It may sometimes happen that the ground becomes so rich in nitrogen (and phosphates) that practically no plants can survive, and some interesting research work could be provided by a detailed examination of " bird-cliffs." There can, at any rate, be no doubt that quite a number of species are excluded from places where birds habitually congregate.

Another interesting peculiarity of certain cliffs, particularly those in the north-west of Scotland and the west of Ireland, is the presence on them of a number of so-called " arctic-alpine " plants—species usually confined to high mountains in the British Isles, although growing down to sea-level farther north. It is not clear why their normal vertical limit in these islands should be depressed in this way, but it may be connected with the unusual humidity of the air as well as the relatively low mean-temperature. One condition, however, seems necessary for the survival of this group of plants wherever they occur—an open habitat, where there is no serious competition from sward-producing species. Roseroot (*Sedum rosea*) is probably the species most frequently seen, for it is often abundant over long stretches of the north Scottish coast. Less frequent is the charming little moss-campion* (*Silene acaulis*), whose neat cushions become sprinkled with delicate pink flowers in June. On basic rocks I have sometimes seen the beautiful purple saxifrage* (*Saxifraga oppositifolia*) and the mountain avens* (*Dryas octopetala*). Other members of this group which turn up occasionally are the mountain sorrel* (*Oxyria digyna*), the cut-leaved saxifrage* (*Saxifraga hypnoides*) and the cowberry or red whortleberry (*Vaccinium vitis-idaea*). It is interesting to note that the reverse phenomenon—the raising of the vertical limit of maritime plants—can often be observed on mountains in the same main districts. Thrift, sea-campion, both the maritime plantains and sea-spleenwort are often found high up in the mountains of the north-west Highlands, and occasionally on mountains farther south (thrift ascends to over 4000 feet on Ben Nevis).

Cliff-vegetation as a whole is perhaps more varied in composition than any of the other types described in this book. In an attempt to generalise, I am conscious of having selected rather arbitrarily certain species as being the most characteristic. Some of my readers may not consider that these are most typical of the cliffs with which they are especially familiar, and may feel that I have omitted to mention the rare plants which are locally common in their districts. Without making the general description unduly complicated, I think this was unavoidable, and if it should stimulate further observations on this attractive, but much neglected, habitat, it may have served a useful purpose. As to the rarer cliff-plants, many of these are referred to in Chapter 12.

N.B.—Those species marked with an asterisk are illustrated in W. H. Pearsall's *Mountains and Moorlands* in this series.

CHAPTER 10

VEGETATION OF CLIFF-TOPS

From a purely scenic point of view, the cliff-tops round the coast of the British Isles are amongst our most priceless possessions (Pl. V, p. 22). If you like walking, they can provide some of the most exhilarating exercise to be found anywhere in the country. The path along the edge of a cliff is usually on the most delightful short springy turf, and the constantly changing views, as each new headland comes into sight, carry one forward without effort. How different is the ardent botanist's painful progress along a shingle beach or over the loose sand between the dunes! But there is also much of interest here for the botanist, for cliff-tops often carry a rich flora and are the home of many rarities.

The area of unploughed land on the tops of cliffs varies greatly. Sometimes only a comparatively narrow strip of ground remains between the edge of the cliffs and the cultivated fields, and this will probably be ungrazed. Elsewhere, where the strip is broader, cliff-tops are often used for rough grazing, and the vegetation will then be correspondingly modified. But in either case there are usually plenty of flowers to be found, and the number of relatively scarce plants which can survive a certain amount of grazing is surprising. It is, however, the common plants which are responsible for the chief displays of colour on cliff-tops. There is, for instance, often an abundance of gorse—either the common form *Ulex europaeus*, which flowers in the spring, or the dwarf form, *Ulex minor*, which blooms in late summer and autumn. In the West, another autumn-flowering species, *Ulex gallii*, with deeper yellow flowers, is prominent. Besides these, the ubiquitous ragwort (*Senecio jacobaea*) and a variety of different yellow " composites " provide an abundance of gold over much of the year. In many places, heather and heath colour the cliff-tops and

Robert Atkinson

Plate 11   A cliff-wall on Skokholm Island, showing rock sea-spurrey,
Spergularia rupicola ; pennywort, *Umbilicus rupestris ;* and English stonecrop,
Sedum anglicum

headlands in August and September, while in the West Country
primroses abound in the spring, and in May some grassy slopes are
carpeted with bluebells and red campion—a lovely combination of
colours, which outshines the delicate misty blue sometimes produced
by thousands of tiny squills (*Scilla verna*) (Pl. 12, p. 158) peeping out
of the close grass turf elsewhere.

The tops of some cliffs are comparatively flat, but more usually
they slope, often steeply, towards the sea. Sloping cliff-tops are
generally much barer, particularly if there is only a thin layer of soil
over the rocks and they are fully exposed to wind and spray. Many
cliffs, too, are much broken up by streams which have cut a way
through them to the sea, producing little sheltered valleys or combes,
so that a walk along the cliff-edge is something of a switch-back.

The nature of the general vegetation found on the tops of cliffs
varies widely, although it usually takes the form of some sort of grassland or moorland. Certain factors are, however, common to all
localities, the most important being their continual exposure to violent
winds. How great this is clearly depends on their angle of slope and
the direction of the prevailing winds, but in practice it is only in the
valleys carved out by streams that any genuine shelter is available.
For this reason, trees are rarely seen in the open, and any woodland
is usually confined to the sheltered slopes of the combes. Even here
it often assumes a dwarfed form, and may be so battered down by
the wind into a tight shrubby tangle that it is almost easier to clamber
over the top than to get through! On the open slopes blackthorn or
gorse may form a low scrub, generally some distance from the edge
of the cliffs and usually flattened by the wind. It is noticeable that
the scrub is much taller and thicker in protected hollows.

A second common factor is that the whole area is exposed to a
certain amount of sea spray, the quantity depending on the distance
from the cliff-edge, its height above the sea, and the direction of the
prevailing winds. No estimates of the amounts of spray falling during
the year have yet been made, but they must be considerable. None
the less, I think it is doubtful whether this influences the composition
of the vegetation appreciably except near the cliff-edge, or on small
islands and exposed headlands, unless the slope is very steep.

The presence of large numbers of rabbits and the shallowness of
the soil are two other factors which are often much in evidence, and
both play a part in keeping the vegetation short and stunted. As the

area is not under cultivation, little effort is usually made to keep the rabbit population in check, with the result that, even when the land is not officially employed for grazing, most of the plants are much nibbled and the grass is kept short. The soil, too, may often be restricted to a thin layer lying on the rock, although this state of affairs is by no means universal. If the soil is shallow, water is frequently in short supply, and this will be aggravated by the rapid surface-evaporation caused by the strong winds—another reason for the stunted vegetation.

We saw in the previous chapter that quite a large number of inland species often appear on the upper, more broken, portions of cliffs, although they are usually associated with enough maritime species to give the whole vegetation a markedly seaside flavour. Amongst the non-maritime plants are some which always seem to occur more abundantly near the sea, although they are not confined to the coastal belt—fennel (*Foeniculum vulgare*), English stonecrop (*Sedum anglicum*) and buck's-horn plantain (*Plantago coronopus*) are obvious examples. Such plants are often loosely described as " submaritime," but we do not at present know enough about their individual ecological requirements to provide a reason for their apparent preference for the coastal region.

Although there is no single species exclusively confined to cliff-tops, a number of these so-called submaritime plants are particularly characteristic of this habitat. Of these, the most charming is the little vernal squill (*Scilla verna*) (Pl. 12, p. 158), which is sometimes very common in the cliff-grassland in the West. I think this is almost my favourite wild flower; only 3 or 4 inches high, it nevertheless some-times grows so thickly that the grass along the cliff-edges acquires a delicious pale blue sheen when it is in flower in early May. Its close relation, the autumnal squill (*Scilla autumnalis*) is also found in similar places, but is much more local in occurrence. Its blue, however, is more mauve, and I have never seen it growing sufficiently thickly to colour the ground when it is flowering in late summer. Another typical cliff-top plant in the West is the sea-storksbill (*Erodium maritimum*), easily distinguishable from the other two storksbills, which are also commonly found there, by its undivided leaves.

The most characteristic submaritime communities are usually restricted to a comparatively narrow strip of ground along the edge of the cliffs, where the plants of the bare rocks meet the closer inland vegetation developed on the cliff-tops. On some cliffs, the slope

gradually eases off, and it is difficult to say where the cliffs end and
the cliff-grassland begins, particularly if much detritus has fallen from
the top over the edge (Pl. XXVIII, p. 147). On others, where the
rocks are steeper, there may be a definite edge separating one
from the other. In most places, however, it is possible to pick
out a zone where only a very thin layer of soil covers the rocks,
not deep enough to support a thick sward and usually showing
a certain amount of bare ground. A careful count of the
different species found along this strip will generally reveal the
presence of a wide variety of plants. These usually consist of a
mixture of inland plants from the cliff-tops, maritime inhabitants
of the open cliffs, and a sprinkling of " submaritimes." Some years
ago, I made some fairly detailed " counts " of the vegetation along
the cliff-edges of the north Cornish coast and found that the following
species (arranged in approximate order of abundance) were most
frequently seen:

The dominant grass of the cliff-
     tops   ..   *Festuca ovina* or *Agrostis setacea*
thrift   ..   ..   ..   *Armeria maritima*
buck's-horn plantain*   ..   *Plantago coronopus* (Pl. XXXVI, p. 179)
ribwort plantain   ..   ..   *Plantago lanceolata*
English stonecrop*   ..   ..   *Sedum anglicum* (Pl. 14, p. 190)
mouse-ear chickweed*   ..   *Cerastium tetrandrum*
storksbill*   ..   ..   *Erodium cicutarium*
sea-pearlwort*   ..   ..   *Sagina maritima*
bird's-foot trefoil   ..   ..   *Lotus corniculatus*
kidney-vetch   ..   ..   ..   *Anthyllis vulneraria*
     (sometimes in its attractive red variety, var: *coccinea*)
vernal squill*   ..   ..   ..   *Scilla verna*
thyme   ..   ..   ..   *Thymus serpyllum*
stalked scurvy-grass   ..   ..   *Cochlearia danica*
Submaritime species are marked with an asterisk.

Several other cliff-plants, such as sea-campion, sea-spurrey or sea-
beet, turned up occasionally, as well as a wide variety of inland species,
especially members of the *Compositae*. It should be mentioned that the
soil on all these cliffs was uniformly acid (average pH of about 5.6),
and that the main vegetation of the tops was rough grassland, with a
certain amount of heather or gorse scrub.

Although the above list is derived from a relatively small area, judging by my observations elsewhere, I think it is fairly typical of the communities found in similar positions on other cliffs, though naturally the relative frequencies of individual species differ widely from place to place. On chalk or limestone cliffs, a greater number of calcicolous species are usually found, but the mixture of inland and coastal plants is of the same type.

In addition to the submaritime species already mentioned, the following are some of the plants of the same class commonly found on cliff-tops elsewhere:

sea-storksbill	..	..	..	*Erodium maritimum*
musk-storksbill		..	..	*Erodium moschatum*
fennel	..	..	..	..	*Foeniculum vulgare* (Pl. XXXIV, p. 175)
alexanders	..	..	..	*Smyrnium olusatrum* (Pl. IV, p. 15)
sea-carrot	..	..	..	*Daucus gingidium* (Pl. XXIX, p. 162)
(and ordinary wild carrot			*Daucus carota*)
slender-flowered thistle		..	*Carduus tenuiflorus*

On most cliffs, the zone which has just been described is comparatively narrow. With increasing amounts of soil available on the inland side, the sward soon becomes too close for the survival of maritime plants, though thrift may manage to hold its own as an intimate mixture with the grass for quite a distance from the edge. The general composition of the vegetation covering the remainder of the cliff-tops is usually much the same as that found on uncultivated land on similar soil in the immediate neighbourhood. Usually this is some kind of grassland, with perhaps a certain amount of scrub in places. Thus on chalk or limestone, typical basic grassland containing a large variety of different species is generally developed. On more acidic rocks, different types of acidic grassland, grass-heath or moorland are found, often with considerable areas of bracken. Alternatively, if there is plenty of humus, typical heathland may occur. However, beyond the occasional presence of submaritime plants, these areas have nothing to offer the seaside botanist, although they are often excellent places to look for inland rarities (Pl. XXV, p. 130).

In very exposed places, such as small islands, stacks or some headlands, where the whole area is exposed to constant spray and wind, but is not grazed, it is sometimes possible to see what happens to the cliff grassland when it is left entirely alone. It then takes the form of a dense growth of tall grasses, usually dominated by red fescue (*Festuca rubra*) or sheep's fescue (*Festuca ovina*), amongst which a variable number of maritime plants such as thrift, sea-campion, sea-spurrey or sea-beet may be found, as well as a few inland species. When the same vegetation is grazed by sheep or rabbits, it is entirely transformed, and then approximates to that just described for the cliff-edges of the mainland.

The small islands and headlands along the west coast of Ireland, which are similarly exposed to spray but are heavily grazed by sheep or rabbits, show a very different type of vegetation, which R. L. Praeger has called "Plantain-sward." Here we have a very low-growing community, sometimes less than an inch high, which is dominated by sea-plantain (*Plantago maritima*), usually in company with buck's-horn plantain (*Plantago coronopus*) and often with ribwort plantain (*Plantago lanceolata*) as well. These three plantains, each occurring in a dwarf form, cover most of the ground with their little rosettes of leaves, and sometimes form a nearly pure community, though elsewhere other species manage to find a footing between them. We have here an example of a community owing its peculiar character largely to biotic factors, since in the absence of grazing, the plantains, with their rosette habit, would undoubtedly be overwhelmed by tall-growing grasses. Very similar vegetation has recently been described for the cliff-edges between the rocks and the moorland on the island of St. Kilda (Outer Hebrides). This consists of a smooth green turf of sea-plantain and red fescue, just over an inch high, with rosettes of thrift at frequent intervals. There is definite evidence here that sea-plantain can only compete with the fescue when its luxuriance is held in check by sheep.

The presence of large numbers of sea-birds on small islands may alter the cliff-top vegetation considerably, just as it does on the open rocks. Thus at St. Kilda, the plantain-sward was largely replaced by a rank growth of Yorkshire fog (*Holcus lanatus*), sorrel (*Rumex acetosa*), meadow buttercup (*Ranunculus acris*), white clover (*Trifolium repens*) and autumn hawkbit (*Leontodon autumnalis*) in places where large numbers of gulls congregated. Where puffins breed in large numbers, the slopes are so riddled with their burrows that the whole surface becomes unstable. The grass here may be completely destroyed, and be replaced by a community dominated by sorrel, which may attain a height of 2-3 feet, with its rank foliage liberally bespattered with bird guano. All the associated species on St. Kilda—common bent-grass

(*Agrostis tenuis*), chickweed (*Stellaria media*), sea-mayweed (*Matricaria maritima*) and Yorkshire fog—were of very robust growth as a result of the abundant nitrogenous matter and phosphates provided by the guano.

Very few observations on the vegetation of cliff-tops have been published and, like that on the cliffs themselves, it offers a largely unexplored field of study for any botanist who has the mind to tackle it.

CHAPTER II

VEGETATION OF BRACKISH WATER

THE SUBMARITIME plants referred to in the previous chapter are by no means confined to cliff-tops, but may be found equally frequently in hedgerows and on waste ground anywhere within a few miles of the sea. There is, however, another group of submaritime plants which differ widely from these, since they are essentially aquatic species. These can best be seen in brackish ditches or swamps to which the tides have no direct access, but where a certain amount of salt water can penetrate and mix with fresh water.

We have already seen (p. 79) that the highest zone of certain salt-marshes (usually the *Juncetum*) occasionally becomes replaced naturally by a fresh-water swamp, particularly if there is a good supply of fresh water coming in from an inland source. Where this occurs, there is always a zone where purely salt-marsh plants mingle with those characteristic of fresh water, and the soil-water is likely to be slightly brackish. This is one of the places where these sub-maritime aquatic plants may turn up, but as a rule most of the seaside species there consist of genuine halophytes from the adjoining salt-marsh. The swamps which sometimes develop on the flat ground behind shingle beaches or narrow belts of sand-dunes, usually show more typical submaritime species. Similar swamps often occur near the mouth of tidal estuaries of slow-moving rivers. In addition to these, the artificial ditches constructed in many salt-marshes in the past to drain the pastures usually carry slightly brackish water and develop a similar type of swamp vegetation. In all these places the vegetation is likely to be made up of a mixture of fresh-water or marsh plants and submaritime aquatic species. The salt-content of the water probably varies considerably with the season, but it is clear

157

that a number of inland water plants are capable of tolerating the relatively low concentration of salt encountered, though plants typical of really acidic swamps and marshes are usually absent.

The vegetation in these rather varied brackish habitats cannot easily be related to the succession in any one of the main groups we have so far discussed, and it therefore seems best to consider it briefly in a separate chapter. Unfortunately it is yet another type of coastal vegetation that has received little study, doubtless because it is difficult to explore in detail without getting wet or, at any rate, covered with mud. Nor, it must be admitted, does it possess any obvious charm, although it may contain one or two attractive plants. None the less, it could provide a fruitful field of research for a keen botanist who was not daunted by the initial difficulties. As things are, the amount of accurate information available about these habitats is too scanty for it to be worth while attempting a detailed description of the vegetation. My own observations, confined to a few places along the south coast and round the Wash, have been more superficial than I could wish, but may suffice to give a general idea of the sort of communities to be found in these types of habitat. It should, perhaps, be pointed out that the moist " slacks " found in some sand-dune areas (see p. 116) do not come into this category, since the water there is usually fresh. When brackish water is found in association with deposits of sand, percolation or occasional flooding by sea-water has generally taken place.

In my experience, by far the most characteristic plant is the sea club-rush (*Scirpus maritimus*), which may be dominant over considerable areas and is thus often responsible for the general appearance of the vegetation. It is a tall plant, usually 2-3 feet high, with long rather broad leaves growing from its base, and may often be seen fringing the banks of brackish ditches or tidal rivers, as well as in the thicker swamps. Another club-rush, *Schoenoplectus* (*Scirpus*) *tabernaemontani*, has a wide distribution and may be locally common. It looks very like the common bullrush (*Schoenoplectus lacustris*), but is shorter, bluish-green in colour, and does not possess any floating leaves. In Scotland and northern England, *Eleocharis uniglumis* and *Blysmus rufus*, smaller plants belonging to closely allied genera, are locally common in brackish marshes.

As might be expected, numerous sedges can usually be found, many of them common inland species. Typical submaritime sedges

John Markham

Plate 12   Spring squills, *Scilla verna*, in cliff-grassland. Anglesey

are *Carex distans*, *C. extensa*, *C. divisa* and *C. punctata*, although only the first two are at all widely distributed. The former is the commonest, being locally abundant in brackish reed-swamps and sometimes appearing on the higher levels of salt-marshes. More easily recognised, and very characteristic of brackish water is the parsley water-dropwort (*Oenanthe lachenalii*), an umbellifer whose upper leaves are divided into long, slender segments. Another umbelliferous plant, which appears to be markedly submaritime, is the wild celery (*Apium graveolens*), interesting as being the original stock from which our garden celery was produced. It is locally frequent in brackish marshes and ditches, and is easily recognised by its stout, furrowed stems from which numerous small umbels of greenish-white flowers issue at intervals. Brookweed (*Samolus valerandi*) is another marsh plant with a pronounced submaritime distribution. Although frequently seen in brackish ditches, it is also found in running water—particularly where little streams run out on to the shore. But perhaps the most attractive plant to look out for in these habitats is the pretty marsh mallow (*Althaea officinalis*), with its large pink flowers and its soft velvety leaves—now unfortunately less common than it used to be (Pl. XXX, p. 163).

In more open water, various green algae, such as *Vaucheria* and *Enteromorpha* species are likely to be present. Various pondweeds, too, are frequent inhabitants, most of them rather uninteresting-looking little plants with grass-like leaves, such as the fennel-leaved pondweed (*Potomogetum pectinatus*), the tassel-pondweeds (*Ruppia maritima* and *R. spiralis*) and the horned pondweed (*Zannichellia palustris*). More easily identified, and very characteristic also of open water, is one of the more robust-looking water-crowfoots, *Ranunculus baudotii*, whose floating leaves are divided into three definite segments, and which sometimes forms extensive patches on the surface of the water. There are, of course, many rather similar species of water-crowfoots, but this appears to be the only one with markedly submaritime tendencies.

Practically all the plants so far mentioned have some claim to be described as submaritime, and doubtless a number of others could be added. There is, however, a rather wide difference between a near-halophyte like *Scirpus maritimus* and such a plant as brookweed, which has a wide inland distribution. Most of the plants referred to belong to an intermediate type, but they have this in common, that they are more frequent near the sea although they may all be found inland.

A certain number of more definitely halophytic species are sometimes found, though they can hardly be called characteristic. Thus eel-grass (*Zostera*) is found in the Fleet behind Chesil Beach, and the long-leaved scurvy-grass (*Cochlearia anglica*) is decidedly characteristic of muddy banks of tidal estuaries. A number of other typical salt-marsh plants turn up from time to time in brackish marshes, perhaps the sea-rushes (*Juncus maritimus* and *J. gerardi*), the sea-arrowgrass (*Triglochin maritima*) (Pl. XXXIX, p. 206) and the sea-wormwood (*Artemisia maritima*) (Pl. XXXI, p. 166) most frequently.

The rest of the vegetation, and sometimes the bulk of it, is composed of inland fresh-water species which have some tolerance of salt. The most frequent plant is undoubtedly the common reed (*Phragmites communis*), which may become dominant over a large area, as for instance at the west end of the Fleet. Reed canary-grass (*Phalaris arundinacea*) may also become locally dominant in some swamps. The number of other water-plants that may occur must be very large, and I can only give a short list of some I have noticed myself, which seem to appear with some regularity:

yellow water-cress	*Rorippa (Nasturtium) sylvestris*
yellow loosestrife	*Lysimachia vulgaris*
rushes	*Juncus* spp.
reedmaces	*Typha latifolia* and *T. angustifolia*
water-plantain	*Alisma plantago-aquatica*
flowering rush	*Butomus umbellatus*
sedges	*Carex* spp.
water-horsetail	*Equisetum fluviatile (limosum)*

In conclusion, it may be worth while to examine the characteristics of submaritime plants in general. The aquatic species we have just discussed make a fairly compact group and may perhaps be best described as "semi-halophytes"—plants which can grow in slightly saline water, but which cannot tolerate the greater concentrations of salt typical of the frequently submerged regions of salt-marshes and do not usually occur in the rarely inundated upper portions because the water-supply there is insufficient. There is no evidence that any of these species can tolerate more than quite small salinities, and it is clear that many common inland water-plants are equally tolerant. The other plants with a submaritime distribution are more varied, but differ widely from these semi-halophytes. They include all those species, other than those typical of brackish water, which are markedly more abundant in the coastal belt than inland. A number of these have been referred to in Chapters 9 and 10, but there are many other more local plants with a similar distribution such as the wild madder (*Rubia peregrina*), the dwarf rest-harrow (*Ononis reclinata*), various clovers such as *Trifolium suffocatum* and *T. molinerii*, and numerous grasses such as the sea-barley (*Hordeum marinum*), etc. Since the majority of these can be found on sea-cliffs, it is clear that they can tolerate small amounts of salt spray, but it is also evident that many typically inland plants can live quite happily under similar conditions.

From what has been said, there can be little doubt that the salt-factor is not the only one to be considered in accounting for the coastal distribution of either of these groups of plants, although in some cases it may be important in suppressing competition by less tolerant species. Without knowing more about the requirements of individual species it would be unwise to suggest what other factors, climatic or otherwise, are involved, but it is extremely unlikely that the same combination of factors is important in each case. For the present we can do no more than note the peculiar distribution of these plants and leave it at that.

CHAPTER 12

THE FLOWERS OF THE COAST
A Descriptive Guide to the
Principal Species

IN THE preceding chapters reference has been made to a considerable number of coastal plants, but little has been said so far about their individual appearance. In this chapter I have therefore compiled short accounts of the majority of plants found round the coast, giving the principal characters by which they can be recognised, their approximate distribution and uses (if any). I have also tried to indicate their relative importance in the general vegetation. It must be emphasised that these brief notes are not intended to take the place of accurate botanical descriptions, which can easily be obtained from a good flora, but are meant solely to give a general idea of the appearance of each plant. Nevertheless, many of the plants described are sufficiently distinct for identification to be quite easy, particularly when an illustration is provided.

The purely maritime plants form a quite small and compact group, and, even if we include with them the species confined to sand-dunes, the total number is not very great. There are, however, a fairly large number of submaritime plants, which occur more commonly in the coastal belt, but are also found inland. It has been difficult to decide how many of these should be included in this survey, and I have compromised by selecting, perhaps rather arbitrarily, those which seem to me most characteristic. Plants which are equally common inland have been deliberately excluded, unless, like the red fescue, they happen to play an unusually important part in the general vegetation.

A good many of the plants described here are extremely local, but

John Markham

Plate XXIX Sea-carrot, *Daucus gingidium,* in flower on a Cornish cliff-top.

Plate XXX   Marsh mallow, *Althaea officinalis*; a beautiful plant found
locally in brackish marshes.   Dorset.

no attempt has been made to include all the rare species that may be met with somewhere along the coast, particularly those differing only slightly from the commoner species of the same genus. The large number of different species of the genus *Salicornia*, for instance, have been excluded, although some of them are widely distributed.

I have arranged the species in the most familiar systematic order under families, since this chapter is intended to be used for easy reference. English names have been provided in as many cases as possible, but there are a number of rare or inconspicuous plants which have no recognised English equivalent. The Latin names employed are usually those adopted in the recently published "Flora of the British Isles" (Clapham, Tutin and Warburg). Synonyms in common use have been added in most cases.

The number of botanical terms has been cut down to a minimum, but a short glossary has been added at the end to explain those whose use could not be avoided. I have also taken the opportunity of including in this glossary the meaning of various ecological terms employed in the earlier chapters.

PAPAVERACEAE

YELLOW HORNED POPPY, *Glaucium flavum*, (Pl. IX, p. 54). This distinct and showy plant is one of our most attractive coastal species. Its most characteristic habitat is shingle, but it is sometimes found on sand, on debris at the foot of cliffs, or on cliff-tops, etc. Although absent from long stretches, and rare in Scotland, it is distributed round the whole coastline and is occasionally seen some miles inland on chalk or limestone. It is particularly luxuriant on some of the East Anglian beaches and on parts of Chesil Beach.

It has a woody rootstock and may be a perennial in some places, but is usually a biennial. The whole plant has a rather cabbage-like appearance, the thick wavy leaves, which clasp the stem, having a marked bluish-green colour. During the winter these remain in the form of a rosette, but a stout stem, up to 3 ft. high, is sent up during the summer. The stem itself is smooth, but the leaves are covered with stiff hairs. The handsome flowers are of a rich golden-yellow, up to 3 in. in diameter, and the seeds are produced in long thin pods, which are curved and glabrous. The sap is orange-coloured and has a strong unpleasant smell. At one time it was thought to be a remedy for

bruises, and the plant is still sometimes called " bruisewort." Flowers:
June-September.

Robert Bridges has a charming little poem about this species:

> *A poppy grows upon the shore,*
> *Bursts her twin cup in summer late:*
> *Her leaves are glaucous-green and hoar,*
> *Her petals yellow, delicate.*
>
> *Oft to her cousins turns her thought,*
> *In wonder if they care that she*
> *Is fed with spray for dew, and caught*
> *By every gale that sweeps the sea.*
>
> *She has no lovers like the red,*
> *That dances with the noble corn:*
> *Her blossoms on the waves are shed,*
> *Where she stands shivering and forlorn.*

> from *The Shorter Poems of Robert Bridges*
> by permission of the Clarendon Press, Oxford.

CRUCIFERAE

SEA-STOCK, *Matthiola sinuata,*. This rare plant is found on open sand-
dunes or on cliffs at various places along our south-west and west
coasts, but not north of the Welsh coast. It has been introduced
on some of the cliffs of the Lizard peninsula (Cornwall), and is
common on sandy shores in the Mediterranean.

Like the garden stock, it is a biennial, producing a branching stem
up to 2 ft. long, which springs from a rosette. The stem-leaves are
long, narrow and somewhat wavy, the radical leaves being more
deeply toothed. The whole plant is hoary, covered with soft, downy
hairs (Fig. 5(d), p. 43). The large flowers, nearly an inch across, grow
in a raceme and are of an attractive pale-lilac colour. They are
distinctly fragrant, particularly at night. The seeds occur in long
thin pods, which are sometimes nearly 4 in. long. Flowers: June-
August.

QUEEN STOCK, *Matthiola incana.* This is a very rare plant, probably only truly wild on certain cliffs in the Isle of Wight, though perhaps also on some other south coast cliffs. Escapes from cultivation are sometimes found along the shore elsewhere.

It is a perennial with a hard wood stem, dividing near the ground into numerous ascending branches and forming a bush 1-2 ft. high. The leaves occur mostly as rosettes at the extremities of the old branches, and thinly scattered along the new shoots. They are lance-shaped, soft and hoary on both sides, and covered with short curly hairs. The large handsome flowers are borne in a short raceme. They are over an inch across and vary in colour from deep violet to pale lilac. The seeds are produced in long hairless cylindrical pods often over 4 in. long. The whole of the rest of the plant is downy. The general appearance resembles closely that of the Brompton Stock of cottage gardens. Flowers: May-June.

COMMON SCURVY-GRASS, *Cochlearia officinalis* (Pl. XXXII, p. 167). This well-known halophyte is a common inhabitant of salt-marshes and also of rocky cliffs. In the former, it is generally commonest on the middle levels, e.g. in the " general salt-marsh community," but also favours the *Juncetum* higher up, where there is some shade. On cliffs, it seems to prefer the damp and shady spots, and is often particularly luxuriant on " bird-cliffs." It is found round the whole coastline, and is particularly common in Scotland.

It is a perennial, with a thick woody rootstock from which spring numerous trailing or ascending angular stems. The plant varies much in size, from only a few inches high in dry places to well over a foot in damp ground. The shining deep-green leaves are thick and fleshy. They vary greatly in shape, the lower ones being roughly kidney-shaped and having long stalks, while those on the stem are generally stalkless and narrower. The flowers are white, sometimes tinged with purple, and are about ¼ in. across. They occur in short racemes, and when seen in a mass on a cliff or salt-marsh produce attractive snowy patches. The seed-pods are globular. Flowers: April-July (one of the first coastal plants to come into bloom).

The English name originates in the fact that the whole plant contains an oil which has anti-scorbutic properties. In the days of long voyages in sailing ships it was a common practice for sailors to eat the leaves as a protection against scurvy, as is recounted in Captain Cook's diaries.

STALKED SCURVY-GRASS, *Cochlearia danica.* This plant may be found on cliffs, sea-walls, muddy estuaries, and more rarely on sand-dunes. It is very common in the South-West, and along parts of the south coast, but becomes less common farther north and is rare in Scotland.

It is a smaller plant than the common species, and is usually a biennial. The chief differences between them are that the leaves are all stalked and ivy-shaped, the flowers are smaller and more frequently tinged with purple, and the pods are more oval. Flowers: April-July.

LONG-LEAVED SCURVY-GRASS, *Cochlearia anglica.* This plant is most characteristic of muddy estuaries and salt-marshes, but is occasionally seen on cliffs and even on sand. It is locally common in suitable places, chiefly in the South. Rare in Scotland.

It is usually larger and more robust than the common species, growing in denser tufts and having more numerous stems up to a foot in height. The stem leaves are narrow, clasping the stem, and the radical leaves taper gradually into the footstalk. The flowers, too, are larger and may be nearly ½ in. across. The pods are oval-oblong, up to ½ in. long. Flowers: May-July.

SCOTTISH SCURVY-GRASS, *Cochlearia scotica* (*groenlandica*). This rare little arctic plant is almost confined to north Scotland and Ireland, where it is found on small shingle, sand, and occasionally on rocks.

It is usually a biennial, growing in close tufts, and looking like a dwarf form of *Cochlearia officinalis.* The small kidney-shaped leaves are hardly toothed and do not clasp the stem. They are usually stalked, the radical leaves being proportionately longer. The little seed-pods are oval. Flowers: June-September.

WILD CABBAGE, *Brassica oleracea* (Pl. 13, p. 183). This Mediterranean species is found locally on sea-cliffs and rocks chiefly along the south and west coasts, though not north of the Great Orme, where it is abundant. Common garden cabbage seedlings are apt to become established on cliffs and may easily be confused with it, but it is genuinely wild in a number of places along the south coast, particularly on the chalk cliffs of Kent, and near Swanage.

It is either a perennial or a biennial, with a thick woody rootstock from which spring a number of thick, twisted and much-branched

John Markham

Plate XXXI Sea-wormwood, *Artemisia maritima*; a familiar inhabitant of salt-marshes and brackish swamps. Essex.

Plate XXXII   Common scurvy-grass, *Cochlearia officinalis*, on a Cornish cliff.

stems, usually more or less prostrate. The leaves are glaucous and somewhat fleshy, with a waxy coating. They all have wavy margins and are roughly oval to oblong in outline, those on the stem being stalkless, but the lower ones having stalks and sometimes becoming very large. The pale-yellow flowers are nearly an inch across and occur in a long raceme. The thin seed-pods have a short beak at their ends and may be over 2 in. long. Flowers: May-August.

The young leaves make excellent " greens," and all the garden forms of cabbage are descendants of the wild plant.

ISLE OF MAN CABBAGE, *Rhyncosinapis* (*Brassicella*) *monensis*. This is a local plant of sandy shores and open dunes, confined to the north-western coast of England, the south-western coasts of Scotland and, of course, the Isle of Man. It is a west European species.

It has a long woody branched rootstock, from which spring rosettes of deeply cut, pinnatifid, glaucous and glabrous leaves, which are quite unlike those of the previous species. The stems are about 1 ft. long, and are more or less prostrate and leafless. The pale-yellow flowers occur in a short raceme, and are not more than ¾ in. across. The seeds are contained in long thin pods, about 2 in. long, which terminate in a long beak. Flowers : June-August.

Rhyncosinapis wrightii is a very rare plant only found on some of the Lundy Island cliffs. It has larger flowers, with broader petals, larger and less deeply cut leaves, and is hairy.

SEA-KALE, *Crambe maritima*. The most characteristic habitat of this west European species is along the drift-zone on shingle or sandy beaches, although it is sometimes found on cliffs, particularly on chalk. Though relatively rare in the British Isles, it occurs in a number of widely separated localities, including south Scotland. It is probably found most frequently along the eastern half of the south coast.

It is a perennial with a thick and fleshy rootstock, which produces subterranean shoots readily when the plant becomes buried in shingle or sand. The rather thick spreading stems are much-branched, and may reach a length of 2 ft. or more. The leaves are roughly oval, coarsely toothed and wavy. The lower ones are on long stalks and are sometimes very large, resembling those of a cabbage. Those on the stem are few and small. The whole plant is fleshy, glaucous, and glabrous. The rather large white flowers are nearly ½ in. across and grow in crowded clusters, the seed-pods being egg-shaped. Flowers: June-August.

The wild plant is still sometimes eaten as a vegetable, but is also widely cultivated. The young shoots, preferably blanched by excluding the light from them, are boiled and eaten like asparagus.

SEA-ROCKET, *Cakile maritima* (Pl. XVI, p. 83). The typical habitat of this well-known plant is the drift-zone along sandy beaches, but it may sometimes be seen on open sand-dunes and, more rarely, on shingle. It is found round the whole coastline and has a wide distribution in the world, including North America and Australia. Its most frequent associate is the prickly saltwort, and the two species sometimes form a thin, but well-marked, line of vegetation along the top of a beach (Pl. VII, p. 30). It can tolerate a considerable amount of burying by sand, and its creeping roots and semi-prostrate stems enable it to collect enough sand to form tiny dunes.

It is a rather bushy annual, with many straggling and branched stems. Both stems and leaves are quite smooth, the latter usually few in number, fleshy and glaucous. They are roughly oblong in shape, somewhat cut into lobes, the degree of dissection varying widely. It is quite a pretty plant when in bloom, the flowers varying in colour from lilac to white. They grow in flat-topped panicles on stiff stalks and are nearly ½ in. across. The seed-pods are spindle-shaped and have a square section. Flowers: June-August.

SEA-RADISH, *Raphanus maritimus*. This west European or Atlantic species is found along the drift-line on sandy shores and shingle banks, and also occasionally on rocky cliffs in a few localities on the south and west coasts of England, in south-west Scotland (e.g. Ailsa Craig), and in various places along the Irish coast.

It is usually biennial, but may sometimes be perennial. A large rosette of irregularly pinnate radical leaves grows from a thick rootstock, and a stout erect stem, covered with bristles, is sent up to a height of 18 in. to 2 ft. The rather large flowers occur in long sprays, the petals being generally pale-yellow, but sometimes nearly white. The pods are constricted into well-marked segments, each containing one seed, and terminate in a prominent beak. The number of seeds in a pod varies from one to four. Like a number of other coastal plants, it presents an untidy and weather-beaten appearance at the end of the flowering season. Flowers: June-August.

VIOLACEAE

SAND PANSY, *Viola curtisii*. This rather rare plant is found in certain sand-dune areas along the west and south-west coasts, in the Hebrides, and in a number of places in Ireland. Where it is established, it may be quite common, e.g. in Braunton Burrows (Devon).

It is a small perennial, with slender branched stems, mostly semi-prostrate, and often growing under the sand and producing little tufts of leaves at intervals. The flowering stems are rough and angular, varying from 1 to 6 in. in height. The flowers are yellow, and have rather broad spreading petals, a little longer than the sepals, and a long spur. Flowers: May-July.

FRANKENIACEAE

SEA-HEATH, *Frankenia laevis*. This common Mediterranean species is found in certain localities from the Wash southwards, down the East Anglian coast and along the eastern portion of the south coast. It is found in the upper levels of salt-marshes, on stable shingle, and occasionally on cliffs. Although its distribution in Britain is so localised, it is fairly common in some areas in East Anglia.

It is a shrubby perennial, with a woody rootstock dividing into numerous wiry trailing stems, 6 to 9 in. long. These are thickly clothed at their extremities with small oblong leaves, growing either in opposite pairs or in whorls of four. They are thick, somewhat fleshy, deeply furrowed, often tinged with red, and the edges are usually rolled back. The little flowers, resembling those of a small pink, occur singly in the axils of the leaves or the forks of the stem. They are purplish or rose-coloured, and have from four to six petals at the end of a tubular calyx. The fruit is a small 3-angled capsule. Flowers: July-September.

CARYOPHYLLACEAE

SEA-CAMPION, *Silene maritima* (Pl. XXXIII, p. 174). This well-known coastal plant is found in many habitats, but is most characteristic of shingle beaches and cliffs, where it often covers large areas and is a beautiful sight when in bloom. It is sometimes found on sand-dunes, and, in common with several other maritime plants, is occasionally found high up on the tops of some Scottish mountains. It occurs round the whole coastline of Britain.

It closely resembles the common bladder-campion (*Silene cucubalus*) in many of its characters, but is now considered a distinct species and natural hybrids between the two plants are rare. The pair of species and their hybrids have, in fact, been the subject of much intensive study from a genetic standpoint in recent years by E. M. Marsden-Jones and W. B. Turrill. They reached the conclusion that the sea-campion survived the Ice Age in Britain, but that *Silene cucubalus* had appeared subsequently.

It is a perennial with an extensive, somewhat woody, branched rootstock, from which a large number of shoots spread along the ground to form a thick cushion. Its low mat-form is admirably suited to its requirements for survival when exposed to powerful winds on cliffs and beaches, and it readily forms new shoots when it becomes partly buried by shifting shingle or sand. It has a powerful effect in stabilising mobile shingle and can also endure considerable amounts of spray. The leaves, rather thick and fleshy, are roughly oblong, bluish-green in colour and glabrous. The erect flowering stems rise from the middle of the mat, usually about 9 in. in height. The flowers are generally solitary, considerably larger and more erect than those of the inland plant, with the petals more deeply cleft. The seed is produced in a round capsule, which splits at the top into six teeth, and is light enough to be dispersed by the wind before it has released all its seeds. Its main flowering season is from April to July, but I have found its flowers blooming in Cornwall from the end of February until late in December.

SEA-SANDWORT, *Honckenya* (*Arenaria*) *peploides*. Another name for this plant is sea-purslane, but as this is also in use for *Halimione* (*Obione*) *portulacoides*, it seems better to avoid confusion by employing the above alternative. It is often one of the first colonists of open sand-dunes and is also common on mobile shingle. In addition, it is a frequent inhabitant of foreshore communities along the drift-zone on sandy beaches. It is generally distributed round the coast, although it is absent from long stretches.

It has a creeping, much-branched, perennial rootstock, which is surprisingly extensive for such a small plant. A mass of largely prostrate branched stems spread out from this, ascending a few inches high at their extremities. It multiplies largely by vegetative growth, and often covers considerable areas with its neat mats. Although it is not a true halophyte, it can survive prolonged exposure to salt spray, and, like the sea-campion, readily forms new shoots when partially buried by sand or shingle. It is thus an effective stabiliser of loose sand or shingle. The small, succulent leaves are oval, stalkless, and arranged in crowded opposite rows down the stems. They are dark-green in colour and have a shiny surface. The inconspicuous little greenish-white flowers occur singly or in clusters of two or three in the forks of the stem and leaf axils, but only open in bright sunshine. Flowers: May-September.

The plant has a rather pungent taste, and is said to be still used as a pickle in Yorkshire.

SEA-PEARLWORT, *Sagina maritima*. This little plant is very similar to the common inland pearlworts, and, though often overlooked, is widely distributed round the whole coastline. It may be found on rocky cliffs, on cliff-tops, on fixed sand-dunes and stable shingle, but always in dry places.

A number of weak semi-trailing stems spring from a rather ill-defined central rosette. The leaves are thick and blunt-ended, somewhat fleshy and quite glabrous. The small flowers grow on erect stalks, with four minute petals or none at all, the sepals in any case always being more prominent. The plant often acquires a purplish tinge, particularly the sepals. Like the other pearlworts, it is an annual. Flowers: June-September.

Inland species, such as *Sagina procumbens* and *S. apetala*, are also common in dry places along the coast.

SEA-SPURREYS, *Spergularia* spp. The three common maritime species of this genus are very similar and are best considered together. They all have much-branched semi-trailing stems and slender, rather fleshy little leaves. These are arranged in pairs at the joints, but owing to the frequent presence of small leaf-shoots in the axils, they appear to be in whorls. The small rosy-pink flowers have 5 petals and occur in little clusters.

172   FLOWERS OF THE COAST

Spergularia salina is a common inhabitant of the middle levels in salt-marshes, a typical member of the " general salt-marsh community," and may occasionally be found on rocks. It is an annual or biennial, with a more or less prostrate stem. The leaves are half-cylindrical, somewhat pointed at their ends and glabrous. The rather deep pink flowers are about ⅓ in. across, the petals being a little shorter than the sepals. Flowers: June-September.

Spergularia marginata is found in similar positions, but is probably less common, although it is widely distributed. It differs in being a larger, more erect plant, and possessing a perennial woody rootstock. The fleshy leaves are glabrous, about an inch long, with blunt ends. The flowers are usually paler and have a white centre, ½-⅔ in. across, the petals being as long as the sepals. Flowers: June-September.

Spergularia rupicola seems to be confined to rocks and cliffs, and has been referred to as " rock sea-spurrey " in this book (Pl. 11, p. 151). It is more local in distribution, although it is common on many cliffs along the west and south coasts. It has a thick perennial rootstock, but the stem, leaves and calyx are all slightly downy. The semi-cylindrical leaves terminate in a fine point. The flowers resemble those of *S. marginata* in size and colour, but lack the white centre. The petals are usually slightly longer than the sepals. Flowers: June-September.

TAMARICACEAE

TAMARISK, *Tamarix gallica* (Pl. 8, p. 126). This familiar seaside plant is a naturalised alien from the Mediterranean. Since it can withstand violent exposure better than any other shrub, it has been much planted as a stabiliser of shingle and sand, and for making hedges, particularly in the southern half of the country.

It is an evergreen shrub, which sometimes attains the stature of a small tree, up to 15 feet in height. In very exposed places it may become much distorted by the wind. The feathery thread-like leaves are very small and of a bright green colour—an example of " leaf reduction " to reduce excessive transpiration. Its delicate foliage looks particularly beautiful when it has become covered with dew or hoar frost. The small pink flowers occur in short crowded spikes, the petals persisting until the fruit ripens. Flowers: May-October.

MALVACEAE

MARSH MALLOW, *Althaea officinalis* (Pl. XXX, p. 163). This local submaritime plant is found in brackish swamps and damp meadows near the sea, but seems to be getting more rare through drainage operations. It is one of the few attractive-looking plants in this rather dismal vegetation. Crabbe refers to it as " *the soft slimy mallow of the marsh.*" It is still locally frequent in the South, but is scarce north of Lincolnshire and Wales.

It has a thick, fleshy perennial rootstock, and produces an erect, slightly branched, round stem, usually 2-4 feet high. Both leaves and stem are covered with a velvet-like felt of soft hairs, the former being roughly heart-shaped and toothed. The flowers resemble those of the common mallow (*Malva sylvestris*), 1½-2 in. across, pale rose in colour, and set on short stalks in the upper axils. There are from six to nine prominent bracts, joined together at the base of the calyx. The carpels are arranged in a ring, as in the garden hollyhock (*Althaea rosea*). Flowers: August-September.

TREE-MALLOW, *Lavatera arborea*. This local plant is found on rocky cliffs, chiefly along the south and west coasts. It is perhaps most frequent in Devon and Cornwall, but elsewhere it is often difficult to decide whether it is native, since it is a common garden plant. It is widely distributed in the Mediterranean region, and is one of the most distinctive and beautiful of coastal flowers.

It is a tall biennial, with a thick woody stem, sometimes as much as 5 feet high, from which grow vigorous annual flowering branches. The roundish, softly-hairy leaves are cut into lobes like those of the common mallow, and grow on long stalks. The large handsome flowers are 1½ in. across, and grow in clusters in a narrow panicle. The petals have a glossy surface, and are of a pale-purple colour with darker centres. There are three distinct large bracts at the base of the calyx. The ripe carpels form a ring round the flower-stalk and look rather like caterpillars. Flowers: July-September.

Lavatera cretica, another Mediterranean species, resembling the common mallow (*Malva sylvestris*), is naturalised in the Scilly Isles and Channel Islands.

GERANIACEAE

PURPLE CRANESBILL, *Geranium purpureum*. This plant is confined to certain shingle beaches along the south coast. It is particularly common on some of the Sussex beaches and on Chesil Bank, where it is one of the most abundant species on the " terrace " and " back." It is an annual or biennial, closely resembling the common herb-robert (*Geranium robertianum*), which is itself commonly seen on stable shingle. It is distinguished from it by its smaller flowers, which are usually more purplish-red in colour, and the anthers, which are yellow and not orange. In addition, the leaves are thicker, more finely divided and nearly hairless. As a rule, it is a smaller plant, but in exposed dry places, herb-robert is often stunted. Flowers: May-August.

COMMON STORKSBILL, *Erodium cicutarium* (agg.). Although by no means confined to the coast, this species is decidedly more common near the sea. It is particularly characteristic of the short turf on cliff-tops, and is also common on fixed dunes. It can be found in dry places along most of the coastline, but is less common in the North. A number of sub-species have been described.

It is usually a biennial, though some of the maritime sub-species produce a thick tap-root, suggesting a longer period of growth. The much-dissected leaves, growing on long stalks, form a compact rosette when the plant is not in flower. The flower-stems vary in height from an inch or two in dry situations to well over 6 in. when the conditions are especially favourable. The whole plant is more or less covered with spreading hairs. The pretty little pink or purple flowers occur in a small loose umbel, and the carpels are attached to a long beak, as in the two succeeding species. Flowers: April-September.

SEA-STORKSBILL, *Erodium maritimum*. The most typical habitat of this rather local species is cliff-grassland, where the sward is thin, but it may sometimes be found on fixed dunes and waste ground near the sea. It is an Atlantic plant, locally common along the west coast, occasionally seen on the south coast, and it has been reported from a few inland stations.

It is an annual or biennial, smaller than the common storksbill,

J. Peterson

Plate XXXIII   Sea-campion, *Silene maritima*, growing on a cliff.   Island of Foula, Shetland.

Plate XXXIV Fennel, *Foeniculum vulgare*; a submaritime plant of
cliff-tops and waysides near the sea. Norfolk.

from which it can readily be distinguished by its leaves, which are not divided, but are roughly oval with rounded lobes along their margins. These form a flat rosette from which the flower-stalks rise, bearing only one or two pale pink flowers. The whole plant is covered with soft hairs and feels rather sticky. Flowers: May-September.

MUSK STORKSBILL, *Erodium moschatum*. This is another local species, found on cliff-tops, on stable dunes, and in dry waysides near the sea, chiefly in the South-West and in Ireland. Although not confined to the coastal belt, it is definitely submaritime.

It is a larger and coarser plant than the common storksbill. The stems are usually much thicker and may be as much as a foot long, although generally semi-prostrate. The leaves, on long stalks, consist of a number of distinct leaflets, which are deeply toothed. The flowers are more numerous and vary in colour from bluish-purple to nearly white. The plant has a slight musk-like smell. Flowers: May-September.

PAPILIONACEAE (LEGUMINOSAE)

SEA-PEA, *Lathyrus maritimus*. This local plant is nearly always found on shingle, and only rarely on sand. It is common along Chesil Beach, growing chiefly on the windswept crest, but elsewhere is confined to various places along the south and east coasts, notably near Aldeburgh. It is eagerly devoured by cattle when they stray on to the shingle.

It is a perennial, with an immensely long woody rootstock, usually black in colour. The stout prostrate stems are quadrangular and all grow out from the same point. The rather large oval leaflets are bluish-green and glabrous, arranged in pairs of from three to six. The large, handsome flowers occur in a raceme of 5-10 blossoms, and are purplish-red when freshly out, fading later to blue. The seed-pods are slightly hairy, 1-2 in. long and not curved. Flowers: July-August.

YELLOW VETCH, *Vicia lutea*. This is a rather rare plant, but it is widely distributed. It is most frequent along parts of the south coast, but has also been found in one or two places in Scotland, and has been introduced near certain ports. It is generally found on shingle, but occurs sometimes on dry cliff-tops and waste ground near the sea.

It is an annual, growing in tufts and producing semi-prostrate branched stems. The leaves are similar to those of the common vetch (*Vicia angustifolia*), but are greyish-green, with rather long and distant hairs. The solitary flowers are long and narrow, usually straw-coloured and tinged with purple, but sometimes entirely purple. The seed-pods are nearly black, and have a long curved beak. Flowers: June-August.

CRASSULACEAE

ENGLISH STONECROP, *Sedum anglicum* (Pl. 14, p. 190). Although by no means confined to the coastal belt, this species is very characteristic of dry cliff-tops and the upper faces of sea-cliffs (usually on acid soil). It is abundant in suitable places along the whole of the west coast, but is less common on the south coast and rare in the East.

It is a small perennial, only a few inches high, but often forming a thick mat or cushion. The stems are much-branched at the base, and more or less prostrate. They are crowded with small, nearly globular leaves, which are shiny and succulent. These often acquire a reddish tinge. The flowers occur in small clusters at the tips of the flowering stems. The pointed petals are white, generally tinged with pink, and are more than twice as long as the short green sepals. Flowers: May-August.

The yellow biting stonecrop (*Sedum acre*) is commonly found on shingle beaches and often on sand-dunes also, but it is equally common inland. When large masses are in flower on dunes or shingle they make gorgeous patches of gold.

MESEMBRYACEAE

HOTTENTOT'S FIG, *Carpobrotus* (*Mesembryanthemum*) *edulis*. This South African plant is now firmly established on cliffs and sea-walls at a number of points along the south coast, particularly in Devon and Cornwall. Various other species of the same genus turn up from time to time as garden escapes. Its native habitat is bare sandy ground, not necessarily coastal.

It is usually seen in large mats, its long robust and angular stems trailing downwards over the surface of the rocks. The leaves, which may be 3-4 in. long, are thick and have a triangular section. They have smooth margins and are quite glabrous. The large flowers are terminal, and possess many rows of petals. The most usual colour is a rather garish magenta, though the native plant usually has yellow flowers. The South African name refers to the edible fruit it produces, but in this country it is usually referred to by its old Latin name— " Mesembryanthemum." Flowers: June-August.

UMBELLIFERAE

SEA-HOLLY, *Eryngium maritimum* (Pl. 1, p. 35). This unmistakable xerophyte is usually found in mobile sand-dunes, and more rarely on shingle. It is thinly distributed round the whole coastline, being rather commoner in the South, but often absent from long stretches.

It is a perennial, with very long creeping roots, sometimes as much as 8 feet in length. Like many dune-plants, it is capable of upward growth through loose sand, should it become buried. The stem is stiff and rigid, about 2 feet high, and usually much-branched. The leaves also are stiff and leathery, possessing a tough spiny edge and a thick waxy cuticle. They are bluish-green in colour, the veins in particular being markedly blue, and clasp the stem. The small blue flowers are tightly packed into globular heads, reminiscent of those of a teasel, with a whorl of spiny bracts below them. Despite its stiff bearing, it is a distinctly handsome plant. Flowers: July-August.

ALEXANDERS, *Smyrnium olusatrum* (Pl. IV, p. 15.) This is a typical submaritime plant, rarely found far from the sea, but not confined to the coast itself. It was formerly cultivated, the stems being eaten like celery, and it is therefore difficult to be sure whether it is truly wild. It is a common Mediterranean plant, and is locally abundant in the British Isles, particularly in the South-West, being found on cliff-tops and in waysides and hedgerows near the coast.

It is a biennial, with a thick fleshy root. It has a stout, solid, branched stem, sometimes as much as 3 feet high, but usually less. The large leaves are shiny and divided into leaflets, which are further subdivided into trefoils. The final segments are oval, coarsely-toothed, and usually of a yellowish-green colour. The leaves are joined to the stem by a flat sheaf. The greenish-yellow flowers are produced in large compound umbels, which are definitely convex. The globular fruits are black, with prominent ridges. The whole plant has a somewhat coarse appearance, and is usually the earliest umbellifer to come into flower. Flowers: April-June.

FENNEL, *Foeniculum vulgare* (Pl. XXXIV, p. 175). This submaritime plant is found on sea-cliffs above the direct spray zone, and may also be seen in hedgerows and waysides in the coastal belt. It is locally common in the southern half of the country, but may frequently have originated as a garden escape.

It has a perennial thickened root, from which an erect, round, shiny, and jointed stem rises. It may reach as much as 4 feet in height in protected places. The large deep-green leaves are smooth and glabrous, and are made up of very narrow, feathery segments. The small yellow flowers occur in umbels with long stalks, combined into a rather large compound umbel. The oval fruits have prominent ridges. The whole plant has a strong characteristic smell, which is not particularly pleasant, but its large feathery leaves make it distinctly elegant. Flowers: July-August.

It has been in cultivation a long time, for use as a flavouring, and also for garnishing. It was formerly a custom in some places to hang some plants over one's front door at midsummer as a deterrent against witchcraft. Falstaff, in *Henry IV*, Part 2, describing Poins, says: " *he plays quoits well, and eats conger and fennel* "—this habit being presumably a sign of manliness!

SAMPHIRE, *Crithmum maritimum* (Pl. 10, p. 142). This is one of the few coastal plants which is almost entirely confined to cliffs. It is said to occur sometimes on sand-dunes, and I have found it on shingle, but usually near rocks. It is a southern European species, often abundant on the cliffs of the west and south coasts, but not seen north of Suffolk on the east coast, or north of the Solway on the west. In places where it occurs in quantity, it may be one of the principal ingredients of the lowest zone of cliff-vegetation, which is fully exposed to spray.

It is a perennial with a long woody rootstock, which penetrates deeply into the crevices in the rocks. The much-branched stem is smooth and solid, up to a foot high, but usually bent in all directions. The smooth bright-green leaves are divided several times into long narrow leaflets. The flowers are greenish-yellow and grow in small, stiff umbels, which are united into fairly large flat-topped compound

Plate XXXV   Sea-mayweed, *Matricaria maritima*;   widespread on shingle
and rocky cliffs.   Cornwall.

Plate XXXVI Buck's-horn plantain, *Plantago coronopus*; a common submaritime plant of cliffs, dunes, etc.

umbels. The fruits are oblong, rather like those of barley. The whole plant has a characteristic smell and taste. Flowers: July-September. The leaves were formerly much gathered, and made into a strong aromatic pickle, by boiling with vinegar and spice. Samphire is still collected for this purpose in some places. There is a reference to this in the well known description of the cliffs near Dover by Edgar in *King Lear*: "*Half-way down hangs one that gathers samphire—dreadful trade! Methinks he seems no bigger than his head.*"

PARSLEY WATER-DROPWORT, *Oenanthe lachenalii*. There are several rather similar species of this genus in Britain, but this one is markedly submaritime, being a characteristic inhabitant of brackish ditches and swamps, although sometimes occurring in fresh-water marshes especially if they are alkaline, as at Wicken Fen. It is widely distributed in England, though never common, and is rare in Scotland.

It is a perennial, the roots consisting of several fleshy fibres, rather than a taproot. The stems are erect, 2 feet or more in height, and furrowed. The upper and lower leaves differ considerably, the segments of the stem-leaves being long and thin, but those of the more divided lower ones being elliptical and much smaller. The white flowers grow in compound umbels consisting of rather few partial umbels, which become hemispherical when in fruit. Flowers: July-September.

SCOTTISH LOVAGE, *Ligusticum scoticum*. This plant is largely confined to sea-cliffs, and occupies rather the same position in Scotland as samphire does in the South. It grows abundantly on many Scottish cliffs and on those of the Hebrides, but in England is confined to the Northumberland coast. It is definitely a northern plant, and is common all round the Arctic Circle.

It is a perennial with a stout branched taproot. It has a thick erect stem, from 9 in. to over 2 feet high, deeply furrowed, hollow, and often tinged with purple. The deeply divided leaves are dark-green and shining, each branch bearing three oval leaflets on a broad stalk. The edges of these are deeply toothed. The flowers are white, sometimes tinged with pink, and occur in compound umbels with prominent whorls of pointed bracts beneath them. The elliptical fruits have prominent ridges. The whole plant is glabrous, and possesses a strong aromatic smell. Flowers: July-September.

F.C.   N

The leaves were formerly eaten as a green vegetable under the name of " shemis."

SEA-CARROT, *Daucus gingidium* (*gummifer*) (Pl. XXIX, p. 162). This Atlantic species closely resembles the common wild carrot (*Daucus carota*), which is also particularly common near the sea in the South. It is, however, more definitely submaritime, being found mainly on the upper portions of cliffs, cliff-tops, and stable sand-dunes. It is quite common in the South-West and along the south coast.

It is generally a biennial, producing a taproot, and an erect stem up to 3 feet in height. The leaves are less deeply divided than those of the common species, and are usually distinctly fleshy. The most obvious difference is that the umbels in this species are either quite flat or convex, while those of the common wild carrot are concave. When both plants grow close to each other a number of intermediate forms can usually be found, and in any case it is not a very distinct species. Flowers: June-August.

RUBIACEAE

WILD MADDER, *Rubia peregrina*. This submaritime plant is largely confined to the south and west coasts of England and Ireland, where it is locally frequent on cliff-tops and in bushy waste places near the sea. It seems to prefer calcareous soil, and has been recorded from Dovedale. In Ireland, however, it does not appear to be noticeably calcicolous.

It is a perennial with weak straggling stems, 2-3 feet long, often sprawling over bushes and clinging to them by the prickles on its leaves and stems. The dark evergreen leaves are glabrous and shiny, growing in whorls of from four to six. They are lanceolate, tapering at the base into very short stalks, over an inch long, and have little prickly teeth along their edges. The yellowish-green flowers are similar in appearance to those of other bedstraws, and the fruit takes the form of a soft 2-lobed black berry. Flowers: June-August.

COMPOSITAE

SEA-ASTER, *Aster tripolium*. This well-known halophyte is an almost universal inhabitant of salt-marshes all round the coast, and is also sometimes seen on rocky cliffs and in inland saline areas. It is particularly abundant on the east coast, but becomes less so in the northern half of the country and in Scotland. In salt-marshes it is generally most abundant in the lower levels, associated with glass-wort and annual seablite, or sometimes co-dominant with sea manna-grass. It is less common in the middle levels, but may become once more frequent in the upper sea-rush zone. It appears to flourish best where the salt-content is not too high, and can be grown in non-saline soils.

It is a handsome plant and is obviously a close relation of the garden " Michaelmas daisy." It is a perennial, growing from a spindle-shaped root, and producing a stout much-branched stem, generally about 2 feet high, but sometimes taller. It has rather large fleshy, glabrous lance-shaped leaves. The numerous flower-heads occur as a compact cluster, the flowers being about ½ in. across, and consisting of bright-yellow discs surrounded with blue or lilac ray florets. Rabbits are particularly fond of nibbling it, so the flowers are sometimes completely absent. It should be in flower during August and September.

There is a well-known variety, var: *discoideus* (Pl. 4, p. 78) without any of the blue ray-florets, which is especially common in East Anglia, but is largely confined to the southern half of the country.

Another variety, var: *crassus*, has larger leaves and a more stumpy stem, and is found on certain cliffs in Cornwall, Pembrokeshire and elsewhere.

GOLDEN SAMPHIRE, *Inula crithmoides* (Pl. 7, p. 119). This handsome plant is found on rocky cliffs, on shingle beaches, and in salt-marshes. It is locally common along the west and south coasts, and on parts of the east coast as far north as the Wash. In Scotland it is very rare. It is a common Mediterranean species. On cliffs, it grows well down to the spray-zone, and when seen in a mass on rocks or shingle makes a brilliant splash of colour.

It is a perennial, with a long woody rootstock. The stems are erect, a foot or more high, densely leafy at the top but often bare of leaves at the base. The long linear leaves are thick and succulent, narrowed to a stalkless base, and strikingly yellowish-green in the distance. The stem is somewhat branched at the top, each branch ending in a solitary flower-head. These are large " daisy-like " blooms, about an inch across, the rays being a golden-yellow and the disc a deeper orange. Flowers: July-August.

SEA-MAYWEED, *Matricaria maritima* (Pl. XXXV, p. 178). This species is very characteristic of cliffs along the west coast, including Scotland. It is also a common shingle plant and may sometimes be found on sandy foreshores. It can grow in the lowest and most exposed zones on cliffs, and has also been shown to be one of the most tolerant species towards the deposits of bird guano found near breeding colonies. It is generally distributed, rather scarce on the east coast, but locally common elsewhere.

It differs only slightly from the common scentless mayweed (*Matricaria inodora*). It is a perennial, with a woody branched rootstock. The leaf segments are distinctly shorter, and usually decidedly succulent. They are deeper green, and have a shiny surface. It generally has fewer flower-heads, and is a smaller and less branched plant. Flowers: June-September.

SEA-WORMWOOD, *Artemisia maritima* (Pl. XXXI, p. 166). This plant is most characteristic of the upper levels of salt-marshes, but may also be found on shingle beaches, particularly when they adjoin areas of salt-marsh. It sometimes grows on cliffs, and is liable to occur in brackish swamps or on any waste ground near the sea. Round most of the coast it is locally common, but is rare in Scotland.

It is a perennial, with a woody branched rootstock from which ascending or semi-prostrate stems grow, a foot or more long. These are generally curved, somewhat woody near the base, and covered with woolly hairs. The leaves are cut into very slender, blunt segments, the undersides being white and woolly, giving the plant an unmistakable silvery appearance in the distance. The small flower-heads are crowded into short spikes, and are usually of a reddish-yellow colour. The whole plant has a strong aromatic smell. Flowers: August-September.

The common wormwood (*Artemisia absinthium*), with silky, not woolly, leaves, stems up to 3 feet in height, and larger dull-yellow flowers, is also commonly found near the coast. Although often seen on inland waste ground, it is more likely to be native in coastal regions. Like the previous species, it is very aromatic, and it is this plant which is employed in making absinthe in France. There is an old rhyme, referring to another of its former uses:

Plate 13   Wild cabbage, *Brassica oleracea,* on a Dorset cliff

*Where chamber is sweeped, and wormwood is strowne
No flea for his life dare abide to be knowne.*

SLENDER-FLOWERED THISTLE, *Carduus tenuiflorus*. Although many different thistles are commonly seen along the coast, particularly on sand-dunes, the above species is definitely submaritime, being found on dunes, cliff-tops, hedgerows and waste places within a few miles of the sea, but only rarely farther inland. It is locally common, though absent from long stretches of the coastline.

It is an annual or biennial, with an erect stem, 2-4 feet high, branching into a slender panicle at the top. The very prickly, lobed leaves are continued down the stem at their bases, and the whole stem has wavy, prickly wings down it, as in the welted thistle (*Carduus crispus*). Both the stem and undersides of the leaves are covered with cottony down. The small oblong or round flower-heads occur in clusters at the top of the stem and branches, the florets being purplish-pink. Flowers: June-August.

Carduus pycnocephalus, which is common on the Continent, is confined to two small areas near Plymouth. It is very similar, but is more slender and the wings on the stems are less prominent. The flower-heads are less numerous and larger, and the leaves are more cottony beneath.

RED-FRUITED DANDELION, *Taraxacum laevigatum (erythrospermum)*. There are several very similar dandelions, and these have been further split up into numerous micro-species. The above species (or group) is fairly well-marked, and deserves a mention here as it is often one of the earliest plants to colonise the bare sand of young sand-dunes, although it is also found in sandy places inland.

It is a perennial, with the usual dandelion tap-root. As a rule it is rather smaller than the common wayside forms, and its leaves are often glaucous and more divided into narrow acute teeth, although they vary greatly in shape (see Fig. 8, p. 46). The fruit is a dark brick-red colour, ending suddenly in a beak, but prickly at the apex. Flowers: April-September.

PLUMBAGINACEAE

THRIFT or SEA-PINK, *Armeria maritima* (Pl. 9, p. 135). There is probably no better-known seaside plant than thrift, and it is certainly one of the most beautiful. It is found equally frequently in salt-marshes, on sea-cliffs and in the turf on cliff-tops, on seaside walls, and on stable shingle. It also sometimes ascends high into mountains[1] and has been found at over 4000 feet on Ben Nevis. It is sometimes dominant in the central levels of a salt-marsh, and is the most constant species in the lowest zones of vegetation on spray-washed cliffs. Although it is definitely a halophyte, it flourishes best when the salt-content is low, and does well in ordinary garden soil.

It is a perennial, with a very long and much-branched woody rootstock. Under favourable conditions it assumes the form of a thick cushion of densely bunched leaves, but in close turf or under grazing it becomes a flat rosette (Fig. 6, p. 45). The leaves are long, narrow and flat, glabrous and rather fleshy. The flower-stalks are leafless, from 6 in. to a foot high, and carry a single head of closely packed rosy flowers, intermixed with chaffy scales or bracts, which form a thick whorl below each head. The individual flowers consist of a funnel-shaped calyx and 5 petals. Both calyx and bracts have a papery appearance, and survive long after the flower has withered. There is also a similar papery sheaf below each flower-head. It may be found in flower from April to October, but its grandest display takes place in May.

COMMON SEA-LAVENDER, *Limonium vulgare* (Pl. 5, p. 87). The sea-lavenders are amongst the most beautiful flowers of the coast. This species is a common inhabitant of the middle levels of salt-marshes, and may be dominant over considerable areas. It appears not to stand up well to grazing, which may account for its relative rarity along the west coast. It is a southern European plant, which does not occur in Ireland, is rare in Scotland and commonest along the east coast of England. When large patches of it are in flower it is a magnificent sight. Although a true halophyte, it can be grown in ordinary garden soil. The gloomy Crabbe does less than justice to this attractive plant when he says:

[1]Though usually in a slightly different form, with a hairy stem and somewhat broader leaves.

" Here on its wiry stem, in rigid bloom,
Grows the salt lavender, that lacks perfume."

It is a perennial, with a thick, black, rather fleshy rootstock. All the leaves are radical, growing either in tufts or rosettes. They are oblong to lance-shaped, narrowed into a short stalk, and glabrous. Some of them may be quite long, up to 4 in. The flower-stalks are without leaves, from 6 in. to a foot or more in height, and bear numerous purplish-blue flowers in a loose, stiff spike. The calyx is also coloured purplish-green, and persists after flowering. There is a green bract under each flower. Flowers: late June-August.

REMOTE-FLOWERED SEA-LAVENDER, *Limonium humile* (*rariflorum*). This species is sometimes found in muddy salt-marshes, generally at the same level as *Limonium vulgare*. It is not nearly as common, but occurs locally down the west coast from South Scotland, along the south coast, and as far as Yorkshire up the east coast. In Ireland it takes the place of common sea-lavender in all salt-marshes.

It closely resembles the common species, but the flower-spikes are not so close together, and the whole inflorescence is more lax and is often incurved. The flower-stems are somewhat angular, from 4 in. to a foot high, and the leaves are narrower and more pointed. Flowers: July-August.

ROCK SEA-LAVENDER, *Limonium binervosum* (Pl. 15, p. 195). This West European species is largely confined to rocky cliffs, but may occasionally be found on shingle and elsewhere. It is one of the plants found growing in the lowest and most exposed zones on cliffs, and can be a lovely sight when a large mass of it is in bloom. It occurs locally southwards from Wigtown on the west coast, along the south coast, and as far north as Lincolnshire on the east coast, but is much commoner in the West.

A perennial with a long woody taproot, it resembles *Limonium vulgare* in many ways. Its leaves are smaller and narrower, with shorter stalks, and have three veins rather than a single one. As a rule, the stems are not so tall and are more branched. The flower-spikes are more compact, but the flowers themselves are somewhat larger and deeper in colour.

MATTED SEA-LAVENDER, *Limonium bellidifolium* (*reticulatum*). This is a much more local species, confined to the eastern counties from south Lincolnshire to Suffolk, although it is common round the Mediterranean. It occurs characteristically on the stable shingle of the lateral hooks attached to the Blakeney and Scolt Head spits, and sometimes turns up on the middle levels of salt-marshes in company with other shingle plants, such as sea-heath and shrubby seablite.

It is a smaller plant than any of the previous species, having the same woody perennial rootstock, but much smaller and more irregular leaves. These are not more than 1 in. long, with short stalks, and grow in a rosette. The flower-stems are semi-prostrate and much branched, the lower branches being usually bare of flowers. The pretty little flowers are pale lilac, with white, membrane-like bracts. Flowers: July-August.

There are several other rare sea-lavenders, such as *Limonium transwallianum*, *L. paradoxum*, and *L. recurvum*, confined to a few special areas in the South and West.

PRIMULACEAE

SEA-MILKWORT, *Glaux maritima*. This delicate little plant is a common inhabitant of salt-marshes, usually on the upper levels and favouring the damper spots. It may also be found sometimes on sandy shores and on rocky cliffs. It is distributed round the whole coastline, although it is by no means present in all salt-marshes. It is a genuine halophyte and is even found near inland saline areas like those at Droitwich (Worcestershire). Another name for it is black saltwort.

It is a perennial, with long, tough stoloniferous roots. There are numerous semi-prostrate, branched stems, rarely more than 6 in. long, with small oval-oblong leaves growing thickly in pairs along them. These are smooth, stalkless, slightly fleshy and glaucous, and may easily be mistaken for those of the sea-sandwort (*Honckenya peploides*). Large numbers of small sessile flowers are produced singly in the axils of most of the leaves. They have no petals, but the calyx is coloured pink instead and is dotted with darker pink spots. It is a charming little plant. Flowers: May-July.

GENTIANACEAE

CENTAURYS, *Centaurium* spp. Although the common centaury, *Centaurium minus* (*umbellatum*), is often found on sand-dunes, dry clifftops, etc. it is equally typical of dry sandy places inland, except in north Scotland and Ireland, where it appears to be definitely submaritime. Various closely allied species, some of which are briefly mentioned below, are always confined to the coastal belt. They are pretty little annuals, whose bright pink flowers close up tightly in wet weather and in early evening.

The common species is usually the largest, growing to as much as a foot in height under favourable conditions, but in dry ground much less. It has an erect square-sectioned stem, usually much-branched in the upper part. The stem leaves are oval-oblong, growing in rather distant pairs, but the radical ones, which are somewhat broader, form a rosette or tuft at the base. All the branches point upwards and end in a small cluster of rather large flowers, which are more or less sessile. The calyx is a little shorter than the corolla-tube. Flowers: June-September.

The narrow-leaved centaury (*Centaurium littorale*) is a local dune-plant of the North, not found (except for one locality) south of a line from Anglesey to Norfolk, and commoner on the west coast. It differs from the common species in its very narrow blunt-ended leaves, usually rather thick and fleshy, and in its deeper pink and slightly larger flowers, which are never produced in the same profusion. The calyx is a little longer than the corolla-tube and, like the leaves, is usually rough with small protuberances.

Centaurium pulchellum is another local species, usually found on sand-dunes, but sometimes inland. It varies greatly in form and size, becoming very dwarf in dry situations and nearly as tall as the common species in others. The stem, too, may be either simple or branched. The oval leaves are distinctly pointed, and there is no radical leaf rosette. The flowers are small and have definite stalks, appearing singly in the axils or at the ends of the branches. The corolla-tube is longer than the calyx.

Other rare submaritime species are *Centaurium tenuiflorum* and *C. portense*, the last-named a very distinct perennial with trailing stems, confined to one spot on the Pembrokeshire coast.

BORAGINACEAE

Oyster-plant or Sea-lungwort, *Mertensia maritima*. This very distinct plant is a northern species, spread over Europe, Asia and America, found locally round the Scottish coast, but in England limited to a few places along the west coast as far south as Anglesey, and on the east coast to one isolated colony at Blakeney (which appears to be its most southerly station in Europe). It is usually found as a foreshore plant on shingle.

It is a perennial with a black rootstock, which produces numerous white stolons. The much-branched stems are about 2 feet long and almost completely prostrate. The numerous oval leaves are thick, fleshy and rough, with prominent horny spots on them, and have a rather fishy taste (said to be like oysters). They occur in two rows, the lower with stalks, the upper without. The attractive flowers are rather small, only ¼ in. across and are purplish-pink when fresh, becoming blue later. They occur in forked sprays at the ends of the branches, the stamens protruding beyond the tubes of the corollas. Flowers: May-August.

CONVOLVULACEAE

Sea-bindweed, *Calystegia soldanella* (Pl. 6, p. 94). This beautiful species may be found in mobile sand-dunes and as a foreshore plant on sandy beaches, or more rarely on shingle. It is local in occurrence, but is distributed round most of the coast-line, though rarest on the east coast.

It is a perennial, with long spreading roots, which make it quite an efficient stabiliser of sand and shingle. The short stems, 6 in. to a foot long, are prostrate and rarely twine like those of other species in this family. The rather small kidney-shaped or roundish leaves are thick, fleshy and glabrous, and have a shiny surface. The handsome pink flowers are unexpectedly large, 1½ in. across, and are borne singly on square-sectioned stems 6-9 in. high. The petals usually have yellowish or reddish streaks on them. Quite often the unmistakable leaves are seen, but no flowers can be found. This is usually due to the attention of rabbits, which are very fond of eating the flowers, but do not touch the leaves. Flowers: June-September.

The young shoots were formerly gathered along the south coast and pickled, as a substitute for samphire.

SOLANACEAE

SEA-BITTERSWEET, *Solanum dulcamara* var: *marinum* (Pl. 2a, p. 50). A very well-marked variety of the common bittersweet or woody night-shade is found locally on shingle beaches along the south coast, especially in Sussex and Dorset. It differs from the normal form in having completely prostrate angular stems, which are also shorter and more branched. The leaves are all heart-shaped, markedly fleshy, and have shorter stalks. The young leaves and the stem are usually slightly hairy. It has the same blue flowers with a central cone of yellow anthers as the common form, and produces the same attractive red berries. Flowers: June-September.

PLANTAGINACEAE

BUCK'S-HORN or STAG'S-HORN PLANTAIN, *Plantago coronopus* (Pl. XXXVI, p. 179). Although not confined to the coastal belt, this plant is definitely submaritime, and can tolerate as much salting as most true halophytes. It grows in many coastal habitats, but it is especially characteristic of cliff-grassland, and on exposed headlands and small islands it may become co-dominant with the sea-plantain. It is also found on rocky cliffs (even in the zone exposed to most spray), on mature sand-dunes, and on stable shingle. It even turns up occasionally in the upper levels of salt-marshes if the soil is sandy. It is common round most of the coast, but is rarer in Scotland.

Unlike other plantains, it is biennial. Its cut-leaved rosettes are always unmistakable, but they vary greatly in size. Very dwarf forms, not much over an inch across, may sometimes be found in thin dry soils, but in damp shady places the rosettes may measure nearly a foot in diameter. The flower-spikes also vary considerably in size and form, but are shorter than those of the sea-plantain. When in flower, the yellow anthers are prominent. A number of different varieties have been described, but since its form always varies so widely with the conditions, they are very difficult to distinguish. Flowers: June-August.

SEA-PLANTAIN, *Plantago maritima* (Pl. XXXVII, p. 198). The most characteristic habitat of this well-known halophyte is a salt-marsh. It is generally found in the middle and upper levels, associated with such plants as thrift, sea-lavender, sea manna-grass, etc. and sometimes becomes dominant over considerable areas. It is also a common inhabitant of spray-washed cliffs, and of cliff-grassland. It is the dominant species in the so-called " plantain-sward " found on the tops of small islands and exposed cliffs on the west coast of Ireland and in the Hebrides, when the grasses are restrained by grazing. It sometimes occurs on shingle, and is one of those maritime species which are occasionally found high up in the Scottish mountains, usually by the sides of streams. It is widely distributed in Europe, Asia and North America, and occurs round the whole of our coastline.

It is a perennial, with a branched woody rootstock. It has long narrow fleshy and rather thick leaves, which are rounded on the back and channelled in front. The green flowers are borne on a thin cylindrical spike, rather similar to that of the greater plantain. When in bloom, the yellow anthers on long stalks are very conspicuous. Flowers: June-September.

This plant varies considerably in size and shape over its geographical range and also when it is found in different types of habitat in a small area. These varieties have been the subject of exhaustive inquiries by Dr. J. W. Gregor, who has shown that these slightly different growth-forms remain constant under cultivation in ordinary soil—a clear case of natural selection.

The leaves of the sea-plantain are much relished by sheep, and it has even been cultivated in north Wales as sheep food. Like thrift, it stands up well to natural grazing in a salt-marsh.

CHENOPODIACEAE

SEA-BEET, *Beta maritima* (*vulgaris*) (Pl. XXXVIII, p. 199). This well-known seaside plant is a typical " spray halophyte," being often seen near the driftline on foreshores, particularly on shingle. It is also one of the plants which grow in the lowest zone on spray-washed cliffs. It is unusual to see it growing in a mass, and as a rule large isolated plants will be seen dotted about at considerable intervals. A southern European species, it is distributed around most of the coastline, though

Plate 14   English stonecrop, *Sedum anglicum ;*   a typical submaritime cliff
plant. Cornwall

it is rather rare in Scotland and often absent from long stretches elsewhere.

It is a coarse-looking perennial, with a thick fleshy root, from which spring many straggling angular stems. These may be from 2 to 3 feet long, and generally lie mostly along the ground, the young shoots turning upwards only at their ends, and terminating in long spikes of small green flowers. The radical leaves are usually large, roughly oval in shape and narrowing into a broad stalk. The stem leaves are narrower and smaller, but like the others are of a shining deep-green colour and fleshy. The small flowers occur in clusters of two or three, with a slender bract under each, arranged thickly along the spikes. The fruits are small nuts enclosed in a thin bladder. Flowers: July-September.

The young leaves make excellent spinach, and are still used for this purpose in Ireland. The plant is obviously a close relative of the domestic sugar-beet and mangold-wurzel, which have been evolved from a Mediterranean species. It grows well in ordinary garden soil.

ORACHES, *Atriplex* spp. There are three maritime oraches, and at least two inland species commonly found along the coast. They are rather dingy and untidy looking annuals with male and female flowers on different plants, but are difficult to distinguish as they are all very variable in habit. In each case the small green flowers occur in clusters on spikes at the ends of the branches, the fruits being enclosed in two distinct, enlarged sepals, and the leaves are always to a greater or lesser extent sprinkled with a whitish meal. Their most characteristic habitat is the drift-zone on sandy, muddy or shingly shores, but they are also found on the landward edges of salt-marshes, along the banks of brackish ditches, and sometimes on spray-washed cliffs (particularly *A. glabriuscula*).

Atriplex littoralis, sometimes called the grass-leaved orache, is the most distinct, and is generally distributed along the coast, particularly in salt-marshes and along brackish ditches. It is usually fairly erect in habit, the stems nearly always striped with green or white, and the whole plant thickly covered with meal. The leaves are long and thin, sometimes slightly toothed. The flower-clusters occur rather close together on slender leafless spikes at the ends of the branches. Flowers: July-September.

Atriplex glabriuscula (*babingtonii*) is common all round the coast, particularly in Scotland, where it is much the commonest orache along the edges of sea-lochs. The stems are usually prostrate, ascending only at the ends, and often striped with green or white. The leaves are roughly triangular, mostly opposite, pale-green and very mealy. The flowers occur in rather distant clusters on leafy spikes, or in the axils of the upper leaves. Flowers: July-September.

Atriplex sabulosa (laciniata), sometimes called the frosted sea-orache, is a more local plant of sandy or shingly foreshores, commoner on the east coast than the west. The stems are semi-prostrate, usually reddish in colour and without stripes. The leaves are irregular, but roughly egg-shaped, and are generally coarsely toothed. They have either very short stalks or none at all. The flowers are produced in short spikes in the axils of the branches. The whole plant is covered with white scales, giving it a silvery appearance. Flowers: July-September.

The two common inland oraches, *Atriplex hastata* and *A. patula*, are also frequently found on foreshores, usually adopting a more prostrate habit than when growing inland.

SEA-PURSLANE, *Halimione (Obione) portulacoides* (Pl. 16, p. 210). Although sometimes found on shingle beaches and more rarely on cliffs, this plant is largely confined to salt-marshes. It clearly favours a well-aerated soil, and is seen most characteristically growing along the sides of tidal channels through the *Asteretum* or *Puccinellietum*, where it can find a better drained substratum. John Gilmour describes this well as " ribbon development," but on the east coast it sometimes spreads over large areas of the central marshes, obliterating nearly all other plants and forming an almost pure community (*Halimionetum*) (Pl. XV, p. 82). It is very common along much of the east coast of England, generally distributed along most of the south coast (usually confined to the creek-sides, but occasionally being co-dominant with *Spartina townsendii* in the lower marshes), but is local or absent from long stretches of salt-marsh on the west coast. Its northerly limit is approximately a line between Berwick and the Duddon estuary (roughly the course of the July and August isotherm of 60° F.), though it has been reported from Ayr and Wigtown in South Scotland.

It is a low shrubby perennial, with a branched woody rootstock. The wiry stems are usually 1-2 feet long, lying mostly along the ground with erect branches springing from them at frequent intervals. They are covered with a reddish-grey bark. The small leaves vary from oval to lance-shaped (actually two varieties, var: *latifolia* and var: *angustifolia*, have been described), and are silvery white as a result of a mealy coating of fine white scales on both sides. The small flowers occur in short spikes at the ends of the branches and resemble those of the oraches. Flowers: July-October.

ANNUAL GLASSWORT or MARSH SAMPHIRE, *Salicornia stricta* (*herbacea*) (Pl. XIII, p. 66). Although undistinguished in appearance, this plant is one of the most familiar halophytes, and is often largely instrumental in starting the growth of a salt-marsh by colonising the mud along its edge, which may have been already partly stabilised by pioneer algae. It often forms almost pure communities in this position, and is later associated with such plants as sea-aster, annual seablite, etc. It usually persists in small quantities at all levels, though it is only dominant in the open communities at the lowest levels. Owing to its shallow roots it is liable to be uprooted in places where the tide runs strongly. It is common in salt-marshes round most of the coast, except in the *Spartina* marshes along the south coast.

It is a highly succulent annual with a short root and erect stems from 6 in. to a foot in height. These are made up of a number of joints, each about half an inch long, thickened above and notched where the next joint begins (Fig. 9, p. 49). There are no proper leaves, the leaf-bases only existing in the form of succulent scales along the stems. Branches issue from the joints and terminate in short tapering spikes in which the inconspicuous little flowers are sunk in groups of three. The whole plant is bright-green, sometimes tinged with dull red or yellow. Flowers: August-September.

In the past, it was often collected and burnt on a large scale, its ash being used to provide soda for glass-making. It has also been used for making a pickle, said to be inferior to that made from rock samphire (*Crithmum maritimum*).

A large number of closely allied annual species of *Salicornia*, some of which are widely distributed, are recognised, but they are difficult to distinguish and need not be considered here.

PERENNIAL GLASSWORT, *Salicornia perennis* (*radicans*), is a local plant of salt-marshes and muddy tidal creeks, only common in a few places in south-east England, but occurring along the East Anglian and southern coasts.

It differs from the annual species in having a creeping rootstock from which long, woody, prostrate stems spread, rooting at intervals. Numerous slender branches are sent up from these, a single plant sometimes forming a dwarf shrub as much as a yard across. The spikes are blunt, not tapering as in the annual species, and the whole plant is usually of a redder or browner tint. Flowers: August-September.

SHRUBBY SEABLITE, *Suaeda fruticosa*. This interesting Mediterranean species is decidedly local in the British Isles, and is only abundant on Chesil Beach and along the north Norfolk coast, the latter being its most northerly limit. It is most characteristic of intermittently mobile shingle, but it is also found on salt-marshes and along creeks, where the substratum is shingly or sandy, and occasionally at the base of sand-dunes. Its roots appear to require plenty of aeration, so it is not found on water-logged soil, but always in places where the drainage is good. It is markedly a plant of the drift-line, its position along the Fleet behind Chesil Bank (Pl. XXIII, p. 114) being very characteristic. Incidentally, its seeds are water-borne. On the Continent it is primarily a halophyte, but it has been shown to grow well in ordinary garden-soil, and in the British Isles it is always found in places with a low salt-content.

It is a perennial, sometimes forming a shrub of 3-4 feet in height. It has a thick woody stem, sometimes as much as 2 in. thick, which in mobile shingle is usually prostrate, but may be erect in stable ground. Numerous erect branches spring at intervals from the stem; the readiness with which these leafy shoots can be sent up when the plant has been buried by shingle is fully discussed on page 52 (Fig. 11). The numerous glabrous fleshy leaves are semi-cylindrical, flat on top and convex below, and are crowded along the stem, arranged spirally. They are dark-green or purplish, and remain evergreen in the South, though the upper leaves generally fall late in the year in Norfolk. The small green flowers occur in clusters of 1-3 in the axils of the leaves, but only plants which are several years old bloom. Flowers: July-September.

ANNUAL SEABLITE, *Suaeda maritima* (Pl. 3, p. 71). This uninteresting-looking plant is widely distributed in most of the salt-marshes round the coast, including the small marshes at the heads of sea-lochs in

John Markham

Plate 15   Rock sea-lavender, *Limonium binervosum*, on Portland Bill

Scotland. It is usually commonest in the lower levels of salt-marshes, associated with such plants as sea-aster, sea manna-grass, and annual glasswort, but it can usually be found in the higher levels also. Though never dominant in a zone, it may often be the commonest associate in the *Asteretum*, *Puccinellietum*, or *Salicornietum*. It appears to flourish best where the substratum is somewhat sandy, and is frequently inundated by sea-water. It is a genuine halophyte, and there is evidence that it definitely requires salt.

It varies considerably in height, being sometimes very small, but occasionally attaining a height of 18 in. or more. The stem is smooth and slender, either erect or straggling, and usually branched. The leaves are glabrous and fleshy, of a pale green colour, but often becoming rather attractively tinged with red in the autumn as shown in the photograph. They are semi-cylindrical, like those of the previous species, but thinner and more pointed. The green flowers occur in small clusters in the axils. Flowers: July-October.

PRICKLY SALTWORT, *Salsola kali* (Pl. I, p. 6). This common seaside plant is almost entirely confined to sandy shores, where it may often be seen forming a thin zone of vegetation along the drift-line with sea-rocket and other foreshore plants. Though only an annual, it has some effect in fixing loose sand, and miniature dunes collect round it. Occasional plants may also be found in mobile marram dunes. Crabbe was obviously familiar with the parts played by glasswort and saltwort in colonising mud and sand, for he writes: "*Here sampire* [glasswort] *banks and saltwort bounds the flood.*" It is generally distributed round the sandy parts of the whole coast-line.

It is a spreading annual, with many branched angular, stems up to a foot in length, erect when young, but usually becoming largely prostrate later on. The dark-green leaves are awl-shaped, ending in a stiff short spine, and succulent. Both leaves and stem are hairy, and the plant is a typical xerophyte, besides being something of a halophyte and able to stand periodical immersion by sea-water. It does, in fact, absorb considerable quantities of salt, and was formerly collected and burnt, like glasswort, to provide soda for the manufacture of soap and glass. The inconspicuous greenish or slightly pink flowers occur singly in the leaf axils, the anthers being pale yellow. Flowers: July-September.

F.C.   O

POLYGONACEAE

KNOT-GRASS species, *Polygonum* spp. Various closely allied species of knot-grass are found round the coast, usually in foreshore communities on sand or shingle, but also occasionally on sand-dunes and other waste ground near the sea. Superficially, none of them differs greatly from the common inland weed (*Polygonum aviculare* agg.).

Polygonum raii, named after John Ray, is an annual or biennial found locally round the south and west coasts of England and the south-west of Scotland, but rare on the east coast. Its straggling branched stems are not woody, and its leaves are rather narrow, distinctly pointed, and do not roll up their edges. The small flowers, usually in threes, occur in the leaf axils, the petals being generally green with red or white margins, though sometimes red all over. The fruit is a nut, rather larger than that of the common species, smooth and shiny. Flowers: July-September.

Polygonum maritimum is confined to sandy shores in a few places in the South-West and the Channel Islands. It is a perennial, with a woody base and thicker, more rigid, prostrate stems. The leaves are larger and thicker, often glaucous, and the edges are inrolled. The flowers and fruit are also larger, the nut again being smooth and shiny.

Polygonum littorale is fairly widely distributed on dunes and waste-ground near the sea. It is a sub-species of the common knot-grass, with prostrate stems and rather thick, fleshy leaves. The petals are more or less tinged with crimson, and the shiny nut has a distinctly rough surface.

CURLED DOCK (maritime variety), *Rumex crispus* var: *trigranulatus*. The common form of this dock is one of the most abundant weeds in the country, but the maritime variety is an extremely characteristic shingle plant, and may also be seen on sandy foreshores and sometimes on cliffs. It is widely distributed round the whole coast.

It is a perennial with a thick fleshy rootstock and erect branched stem, up to 4 feet high, and often reddish in colour. The long leaves are lance-shaped, wavy, glabrous, and usually more fleshy than those of the inland form. During much of the year the dead leaves of the previous season protect the rosette of radical leaves from spray. The flowers occur in a panicle in crowded whorls. The variety can be distinguished from the inland form by the presence of a little tubercle on each of the outer perianth segments instead of on just one of them. Flowers: June-August.

Several other docks are found locally along the coast, such as *Rumex maritimus* and *Rumex rupestris*. The latter, a south-west European species, found on rocks and shingle at certain points along the south-western seaboard, resembles the sharp dock (*R. conglomeratus*), but is less leafy and produces a more compact, tapering panicle, confined to the upper part of the stem. A South American species, *Rumex cuneifolius*, is now well established in several sand-dune areas in the South-West.

ELEAGNACEAE

SEA-BUCKTHORN, *Hippophae rhamnoides*. This is the only shrub of any size belonging especially to the coastal belt. In some places on the east coast it is the dominant species in the scrub which ultimately develops on the oldest sand-dunes (Pl. XX, p. 103). Elsewhere, it is sometimes found on cliff-tops, and has also been planted in a number of places as a sand-binder. It is locally common, particularly on the east coast, but is probably a genuine native only from Yorkshire to Sussex. It is sometimes called sallow-thorn or willow-thorn, which are really better names, as its foliage is more like that of a willow than a buckthorn.

It is a compact prickly shrub, sometimes 12 feet or more in height, but generally much less. The branches have a tendency to droop, and many of them, instead of lengthening, harden into long spines like those of a buckthorn. Both the stem and branches are clothed in grey bark. The leaves are long and lance-shaped, greyish-green on top, the undersides being covered with scales (Fig. 5e, p. 43), some of which are silvery-white and others rust-coloured. Later in the season when the flowers are over, the edges of the leaves tend to become rolled back. The small greenish flowers appear on the wood of the preceding year round the base of the new shoots, male and female flowers being on separate plants. The male flowers form little clusters resembling catkins, but the females occur singly in the axils, usually in great numbers. The fruits develop into attractive orange berries in September; these are edible, but are usually patronised only by birds. Flowers: May-July.

198   FLOWERS OF THE COAST

EUPHORBIACEAE

SEA-SPURGE, *Euphorbia paralias* (Pl. XIX, p. 102). This plant is most characteristic of mobile sand-dunes, and may also be found along the drift-line on sandy foreshores with such plants as sea-rocket and prickly saltwort. It is a southern European species, frequent in suitable places down the west coast from Cumberland round to Hampshire, but rare on the east coast and absent from Scotland. It is also frequent on the east coast of Ireland.

It is a perennial with a hard woody rootstock, from which bushy-looking erect and stiff stems rise, up to about a foot in height. Its leaves are leathery and fleshy, glabrous, and of a pale bluish-green, sometimes tinged with red. These are roughly oblong, and are crowded along the stem so that they overlap each other, but the lower part of the stem is often leafless. Only some of the stems produce flowers, which occur in a rather compact umbel, the bracts being round or kidney-shaped. Flowers: July-October.

PORTLAND SPURGE, *Euphorbia portlandica*. This is a more local plant than the last, found in similar places but perhaps more typically on cliff-tops. It is a West European species, but in the British Isles is only found along the western portion of the south coast, and along much of the west coast. It is also frequent along the south and east coasts of Ireland.

It is usually a perennial, but has a slender root and more the habit of an annual. Numerous rather slender stems emerge from the base to make a small bush about 1 foot high. The leaves are rather leathery, but thinner than those of the sea-spurge and end in a fine point. They do not clasp the stem to the same extent, and fall early, so as to leave the lower part of the stem markedly bare. The flower umbels are more spreading and diffuse and the bracts are broadly triangular. It resembles the previous species in that the whole plant often becomes tinged with red in the autumn. Flowers: July-October.

SALICACEAE

CREEPING WILLOW, *Salix repens* subspecies: *argentea*. This dwarf willow is widely distributed on damp heaths and moors, but the sub-species is largely confined to sand-dunes. In some places it plays

Plate XXXVII Sea-plantain, *Plantago maritina*; a common salt-marsh and cliff plant. Cornwall.

Plate XXXVIII Sea-beet, *Beta maritima*; **a** well-known foreshore and cliff-plant.

an important part by colonising the damp "slacks" between ranges of dunes, from which it often spreads to the drier sand of the dunes themselves (Pl. XXII, p. 111). It has considerable powers of growing upwards through mobile sand, and sometimes forms quite extensive secondary dunes in broad hollows where sufficient sand is being blown. Its associates in these places are similar to those of open marram-grass dunes, but in the damp slacks, they are chiefly marsh plants. In the British Isles it is found mainly on west coast sand-dunes, but it is conspicuous on the dunes of north-east France and those along the coast of Belgium.

It is a small perennial shrub, with much-branched semi-prostrate stems, which creep extensively underground. The small leaves, varying in shape from oval to lance-shaped, are always silky on the undersides, the young ones being silky on both sides. The cylindrical catkins are practically stalkless, and open at the same time as the leaf-buds. The male and female flowers occur on separate plants, the yellow stamens of the male flowers being very prominent. The fruits are silky. Flowers: April-May.

LILIACEAE

WILD ASPARAGUS, *Asparagus officinalis* subspecies: *prostratus*. Although it is widely distributed round the south-western coasts of Europe, this species is now very rare in the British Isles, being confined to a few localities in the South-West and West, where it grows on sandy shores or cliffs.

It is a curious plant, with a matted, creeping perennial rootstock from which arise annual branching stems, a foot or more in height and usually semi-prostrate. These carry small triangular scales (the rudimentary leaves), and from the axils of these spring clusters of feathery green "leaves," which are really modified stalks or cladodes (Fig. 7, p. 46)—an example of leaf-reduction discussed on page 44. The small dingy yellow flowers are bell-shaped, occurring singly or in pairs in the axils of the principal branches. The fruit is a small, globular red berry. The seeds of garden asparagus are often distributed by birds and sometimes germinate on sandy shores, the young plants being easily confused with the wild plant. The domestic form is very similar, but produces much taller upright stems, at least 4 feet in height and often more. Asparagus has been cultivated since the time of the Romans, the garden form having been produced from the seaside plant. Flowers: June-July.

SPRING SQUILL, *Scilla verna* (Pl. 12, p. 158). This charming little plant is found most frequently in the short turf on cliff-tops, or along the rocky ledges just below, though it has been found inland. It is a typical Atlantic species, occurring from Norway, through Great Britain to western France and Spain, nearly always in submaritime habitats. It is local in the British Isles, though where well-established it may be found in thousands. It is generally distributed along much of the west coast, including Scotland, but is very rare on the east coast. It is also frequent along the east and north-east coasts of Ireland.

It grows from a small bulb, about the size of a hazel-nut. The slender leaves are narrow and channelled, with a prominent hood at their ends. The flower-stalks are only 4-6 in. tall, and in dry ground are even less. These carry a few (up to a dozen) delicate little bells of pale blue, usually with a dark stripe down the middle. The anthers are purple. On a sunny day, when they are in bloom in a mass, they produce a distinct fragrance, and are a lovely sight. Flowers: April-May.

Syrup of Squills (extracted from the bulbs) used to be employed as a cough cure. Crabbe refers to its use in this way by a quack doctor in his poem " The Borough ":

> " A potent thing, 'twas said to cure the ills
> Of ailing lungs, the Oxymel of Squills.
> Squills he procured, but found the bitter strong
> And most unpleasant, none would take it long."

AUTUMNAL SQUILL, *Scilla autumnalis*. This southern European species is much more local in the British Isles than the vernal squill and is confined to the south and west coasts, but does not ascend far north. As a seaside plant it is commonest in Devon and Cornwall, but is found in some inland districts and is not markedly submaritime elsewhere in Europe.

It is a somewhat larger plant than the previous species, growing from a larger bulb, an inch or more in diameter, and producing flower-stems up to 9 in. The flowers are generally few, varying in colour from purplish-red to nearly white. The chief difference is that the leaves appear as a tuft by the side of the stem when the flowers are out, and are only fully developed in late autumn, when they may become 3-4 in. long. Flowers: August-September.

JUNCACEAE

SEA-RUSH, *Juncus maritimus.* Many salt-marshes have a well-marked zone along their upper edge dominated by this plant (*Juncetum*), representing the final stage in the development of the salt-marsh proper. Owing to its tall growth it tends to destroy the close turf of sea manna-grass and other halophytes typical of the middle levels, with the result that some of the plants which are more characteristic of the lowest levels reappear in this zone, though thrift, sea-milkwort, sea-plantain and other " general salt-marsh " plants are its commonest associates. It also sometimes grows in moist slacks in sand-dunes, and has even been planted as a sand-fixer in some districts, since its long roots form a thick mat. It is generally distributed round the coast, being commonest in East Anglia, and rather rare in Scotland.

It is a perennial, growing in irregular tufts from extensive creeping roots. The rigid stems are hard and wiry, varying from 2 to 4 feet in height, and ending in a sharp point. The leaves are shorter than the stems, but are similarly smooth and rounded and have sharp points. The pale flowers occur in rather dense lateral clusters, overtopped by a long pointed bract, which is really the continuation of the stem. Flowers: July-August.

MUD-RUSH, *Juncus gerardi.* Like the previous species, this rush is sometimes the dominant of a community at the highest level in a salt-marsh. In the West, it is often co-dominant with red fescue, and in the South may perhaps be the normal successor to mature *Spartinetum*, when it is not grazed. Where there is a well-developed sea-rush zone, it is frequently the commonest associate and may even be co-dominant. In addition, it is often seen along the banks of tidal rivers, in brackish swamps, and occasionally in moist slacks among sand-dunes. Although widely distributed round the coast, it is less common than the sea-rush and appears to prefer a muddy sub-stratum.

It is usually a smaller plant with more slender erect stems, varying from 6 in. to 2 feet in height and generally 3-angled near the top. It has a similar perennial creeping rootstock, but the leaves are very narrow, somewhat channelled and shiny, and mostly growing from the base. The brownish flowers do not occur in a compact (apparently) lateral cluster, but in a rather loose panicle at the end of the stems. Flowers: July-August.

GREATER SEA-RUSH, *Juncus acutus*. This rush occurs locally in the southern half of the British Isles, but is not found north of Caernarvon and Norfolk and is commoner on the west coast. It is particularly characteristic of moist slacks in sand-dunes, but may also be found in brackish swamps.

It is a tall plant, growing in large round tufts and producing very rigid stems up to 4 feet in height. It resembles the sea-rush, but the flowers are larger, less numerous and browner. The leaves are round-sectioned and shiny, and have the sharpest points of any rush. It is sometimes called the sharp rush, and perhaps Crabbe had this species in mind when he wrote: " *The rushes sharp, that on the borders grow.*" Flowers: July-August.

JUNCAGINACEAE

SEA-ARROWGRASS, *Triglochin maritima* (Pl. XXXIX p. 195). This common salt-marsh plant is usually found in the middle and upper levels, associated with such plants as sea manna-grass, thrift, sea-plantain, sea-lavender, etc. but never occurring sufficiently thickly to be a dominant. In Scotland it sometimes fringes the drainage channels as sea-purslane does farther south, and elsewhere is commonly found in brackish swamps and ditches. It is a genuine halophyte, and is generally distributed round the coast.

It has a perennial rootstock consisting of numerous slender bulbs growing round a common rhizome. The long, very slender leaves are rounded at the back and concave in front, although the ends are flat. They are decidedly fleshy and grow straight up from the rootstock, sometimes to a length of 18 in. The stout flowering stem, varying from 6 in. to nearly 2 feet in height, is usually curved, the upper third bearing a large number of small greenish-yellow flowers in a long, slender spike. These are slightly larger and closer together than those of the inland marsh species (*Triglochin palustris*), which it closely resembles in many other respects. Flowers: May-September.

ZANNICHELLIACEAE

HORNED PONDWEED, *Zannichellia palustris* (agg.). This is an inconspicuous submerged aquatic plant, with long grass-like leaves similar in shape to those of *Zostera*. Although sometimes found in inland ponds, it is usually submaritime and is generally distributed round the coast in brackish water. Various sub-species have been described.

It is a perennial with very slender, branched stems. The narrow, submerged leaves, 1-3 in. long, are opposite and pale green in colour. The minute flowers occur singly or in pairs in the axils, issuing from a sheaf-like bract. Pollination occurs under water. Flowers: May-August.

The tassel-pondweeds, *Ruppia maritima* and *R. spiralis* (*Ruppiaceae*), also found in brackish water, are very similar but have alternate leaves and flowers borne at the surface on wavy flower-stalks. All these plants are frequently overlooked by amateur botanists but are characteristic inhabitants of brackish water.

ZOSTERACEAE

EEL-GRASS or GRASS-WRACK, *Zostera* spp. Three very similar species of this genus, comprising the only flowering plants that are truly marine, occur in the British Isles. They are usually found rooting in sand or mud in protected bays and estuaries between the high and low-tide marks, although *Zostera marina*, at any rate, continues well below the low-tide level. They are perennials with long grass-like leaves, and have creeping stems which run along the surface, rooting at intervals. The small green flowers are enclosed in a hollowed-out sheaf at the base of the leaves. Pollination takes place under water, the thread-like pollen being carried by the currents to the stigmas.

Zostera marina is probably the commonest, being fairly well distributed along the south and East Anglian coasts, but less common in Wales and north-west England. It may form extensive submarine meadows, usually in the form of pure communities apart from marine algae (*Zosteretum*), but is sometimes associated with tassel-pondweed, e.g. in the Fleet behind Chesil Beach. It prefers firm muddy sand, and may play an important part in stabilising the ground before the arrival of later colonists such as annual glasswort. Of the three, this species is least tolerant of exposure, and is rarely found more than about 12 feet above the low-tide mark.

It is the largest species, the flexible, ribbon-like leaves being usually at least a foot in length and sometimes as much as 3 feet. They are succulent, bright green when kept moist but soon bleaching to pure white when exposed for any considerable period on the shore. Flowers: July-September.

During the thirties a serious disease attacked this plant on both sides of the Atlantic, with the result that 90 per cent of the plants in west Europe and the eastern United States were destroyed, and many mud-banks, largely held together by them, collapsed. Only in the Mediterranean and on the Pacific coast of North America were they unaffected. Since 1934 there seems to have been little change in its abundance; it can still be found in small quantity in most of its old localities, and perhaps in time the former beds will be gradually re-established. Owing to the toughness and flexibility of the dry leaves they were used for a variety of purposes, such as a stuffing for cushions and mattresses, as heat insulators, and for packing china and glass, as well as for use as manure in the same way as seaweed. As a result of the disease its use in these ways has been greatly reduced.

Zostera hornemanniana is very similar, and was described for the first time by T. G. Tutin in 1936. It is rather more tolerant of exposure and occurs at least up to the half-tide mark. Its habitat is the same, but in general it is found on softer bottoms and in shallower water. It appears to be fairly widely distributed, having been overlooked in the past as a mere variety of *Zostera marina*. It is a rather smaller plant, and differs chiefly in possessing narrower leaves which are notched at the apex when mature.

The small eel-grass (*Zostera nana*) is more local, but is found in similar habitats. Of the three, it can stand exposure best and generally occurs between the high and low-water marks. It is a much smaller plant, with unbranched flowering stems and slender leaves, 3-6 in. long. Flowers: April-August.

CYPERACEAE

SEA CLUB-RUSH, *Scirpus maritimus*. This is perhaps the most typical plant of brackish water, being commonly found along the banks of tidal rivers and dykes, or forming a border to saline swamps. It may also appear in the upper levels of salt-marshes, particularly where fresh water from inland streams joins the *Juncetum*. It sometimes dominates a community in succession to *Spartinetum* along the south coast. As it is relished by cattle, it is rarely prominent in heavily grazed marshes. It is generally distributed round the coast in suitable places.

It has a black, extensively creeping, perennial rootstock. The plant is usually 2-3 ft. high, with a stout triangular stem, leafy only in its lower half. The shining leaves are deep-green, up to 2 feet in length. The flowers are borne in a few large brown spikelets, which form a terminal cluster. Flowers: July-September.

Schoenoplectus (Scirpus) tabernaemontani is widely distributed and locally common in similar habitats. It closely resembles the common bulrush (*Schoenoplectus lacustris*) and is often confused with it. It is, however, shorter (1-4 feet) and of a dull bluish-green colour. It has the same smooth cylindrical stems, but there are never any floating leaves. The flowers occur in similar compact clusters at the ends of the stems. Flowers: July-September.

SAND-SEDGE, *Carex arenaria* (Pl. X, p. 55). This common and distinct sedge can be found on sand-dunes nearly everywhere round the coast, and not infrequently on shingle as well. It is also prominent on blown sand in the Breckland and some other inland sandy districts. One of the earliest colonists to arrive on open dunes, it often continues as an important ingredient in the vegetation of mature dunes. It is also one of the principal species responsible for the secondary colonisation of " blow-outs." Its extensive roots give it considerable powers of binding the loose surface-sand and it has sometimes been planted for this purpose.

It is a perennial with long creeping roots, which often spread in straight lines just below the surface of the sand for long distances, their position clearly marked by tufts of leaves and flowering stems at intervals. The leaves are rigid with inrolled margins as in many other xerophytes. The stems vary from 3 in. to nearly a foot high, leafy only where they emerge from the sand, and erect until they have flowered, when they usually become curved. The flowers are borne in a crowded terminal spike, the male flowers above and the female below, but not ripe at the same time to ensure cross-pollination. Flowers: June-July.

A number of more local sedges appear to be distinctly submaritime in distribution, and are characteristic of brackish swamps and the upper portions of salt-marshes. Amongst these are *Carex distans, C. divisa, C. extensa, C. maritima (incurva),* and *C. punctata.*

GRAMINEAE

COMMON CORD-GRASS, *Spartina maritima (stricta).* This is the only native species belonging to this genus, being usually found as an early colonist on the lower levels of muddy salt-marshes and near the mouths of tidal rivers. It is locally common along the south and east coasts, and sometimes becomes dominant over small areas, usually in succession to *Salicornietum* or *Asteretum,* but occasionally colonising the bare mud. It occurs locally along the shores of the Atlantic and is common in South-west Europe.

It is a perennial, with an extensive creeping rootstock, producing numerous tufts. The stems are erect and rigid, from a foot to 18 in. high, and somewhat succulent. The leaves are stiff, smooth and flat when fresh, though the edges roll inwards when they are dry. They are rather short, from 2 to 6 in. long, and are easily detached from the sheafs along the stem, the lower ones always falling early. The flowers usually occur in two spikes, though sometimes there are more, but they are so close together that they appear as a single long spike. Flowers: July-September.

Spartina alterniflora is a rare alien from North America, found on the mud of Southampton Water, and first recorded in 1829. It is chiefly interesting as being a parent of *Spartina townsendii.* A much taller plant than the common cord-grass, it has stems up to 4 feet in height. The leaves are also much longer, up to a foot in length, but not jointed to the sheaves. The flower-spikes may number anything from 5 to 8, and they differ in being longer and more loosely applied to each other, as well as being overtopped by the leaves.

RICE-GRASS, *Spartina townsendii* (Pl. XIVa, p. 67). This remarkable grass is a natural hybrid between *Spartina maritima* and *S. alterniflora,* first reported from Southampton Water about 1870, where the latter had been established for some time. Since then it has spread with astonishing speed and is now abundant in the muddy estuaries and salt-marshes of the south coast from Sussex to Dorset and has appeared in a number

Plate XXXIX Sea-arrowgrass, *Triglochin maritima*; a familiar salt-marsh plant. Norfolk.

Plate XL   Sea-spleenwort, *Asplenium marinum*, our only maritime fern, growing on a wall. Portquin, Cornwall.

of places on the east and west coasts. It flourishes best in soft, sloppy mud and possesses greater powers than any other known plant in stabilising mobile mud (see p. 71). It spreads over new ground with extraordinary vigour, and causes a rapid rise in the general level by collecting new material. It forms almost pure communities along the south coast (*Spartinetum*) (Pl. XIVb, p. 67), and has been successfully planted at a number of points elsewhere, notably on the east side of the Wash. Despite its immense vigour, it has not yet shown any definite tendency to oust *Salicornia* from its position as a primary colonist in those marshes where the substratum is sufficiently firm for it to become established.

It is a much larger plant than *Spartina maritima*, and grows in large circular tufts. It anchors itself in the mud by long, stout, vertically descending roots, while horizontal feeding roots, much interlaced and divided, spread through the surface layers, and stolons radiate in all directions from the bases of the stems (Fig. 9c, p. 49). The stout stems are generally about 2 feet high, but may be considerably taller. The leaf-blades are jointed to the sheafs like those of *S. maritima*, and are similarly rigid and erect, but longer. The stiff and slender flower spikes usually number from 4 to 7, and are fairly closely pressed to each other, overtopping the tallest leaves. Flowers: July-September.

Mature *Spartina* meadows furnish good grazing and the grass is relished by cattle.

SAND CATSTAIL or SAND TIMOTHY-GRASS, *Phleum arenarium*. This small annual grass is not uncommon in sand-dune areas round the coast, usually on partially fixed sand where the surface-cover is not too close. It occurs also in a few inland sandy areas, notably in Breckland. It is rather ephemeral, belonging to the class of " winter annuals," and in late summer when the surface of the sand is dry it is generally quite dried up, its life-cycle completed.

It grows in small tufts, with erect and somewhat branched stems 2-9 in. in height. The leaves are very short, about 1 in. long, flat, slightly ribbed, and bluish-green. The flowers occur in an egg-shaped or cylindrical panicle, slightly tapered below and much shorter in proportion than that of the common timothy-grass (*Phleum pratense*). The glumes are bristly or hairy. It is a distinct little grass, but easily recognisable as a miniature " timothy." Flowers: May-June.

MARRAM-GRASS, *Ammophila arenaria* (Pl. III, p. 14). This is the most familiar and easily recognised of all coastal grasses. It can be found in nearly all places where mobile sand exists, and has been the chief agent in building up the sand-dune areas round the whole of the coast. Its immensely long creeping roots and the vigour with which it grows up through sand when buried make it a most efficient sand-binder. Its action in building up dunes is fully discussed on page 51. It is a plant which definitely prefers mobile sand, for as soon as the surface sand becomes stabilised in the older dunes, it ceases to flower and is eventually choked out. It is thus dominant only in the earlier phases of sand-dune development. It is not a halophyte and cannot stand prolonged immersion by sea-water.

It produces a mass of creeping rhizomes of great length, but in a large dune, only those relatively near the surface are usually still alive. The stems are tall and rigid, from 2 to 3 feet high. The leaves are long and narrow, ending in a sharp point and inrolled like those of many xerophytes, particularly in dry weather. The outer surface is smooth and glossy, of a bluish-green colour. The inner surface has prominent grooves and is hairy, both devices for checking excessive transpiration (Fig. 4, p. 42). The flowers are crowded into a long cylindrical spike-like panicle, about 6 in. long, tapered at both ends. Flowers: July-August.

The long roots have often been used for making mats, hence its alternative name of " mat-grass." The leaves, also, are sometimes used for thatching. It has been planted on a large scale both in this country and abroad as a stabiliser of shifting sand, and it is illegal to remove or destroy the grass in many countries.

Ammocalamagrostis baltica, formerly known as *Ammophila baltica*, is a natural hybrid between marram-grass and the wood smallreed (*Calamagrostis epigejos*), which is established on sand-dunes at a few places on the Suffolk and Northumberland coasts. It is chiefly distinguished by its long thin panicle, which is curved rather than stiffly erect, and is often purplish in colour.

GREY HAIR-GRASS, *Corynephorus canescens*. This rare little grass is confined to a few places in East Anglia and the west coast of Scotland, though it extends over southern and central Europe in sandy places and is by no means confined to the coast there. Amongst other places, it occurs in the sand-dunes at Blakeney (Norfolk), and farther east (at Salthouse) it forms a short turf along the edge of the shingle bank.

It is a perennial, growing in dense little tufts. The stems are slender, not more than 9 in. high and generally less, usually bent near the base. The leaves are needle-like, with inrolled margins, rough, and red or purplish below. The flowers occur in a narrow panicle, which becomes more spreading when the flowers open. The anthers are purple. Flowers: June-July.

SEA MANNA-GRASS, *Puccinellia (Glyceria) maritima.* This is by far the commonest salt-marsh grass, although its importance in the general vegetation varies greatly in different areas. It appears to prefer a sandy substratum, and thus plays a relatively small part in the development of the East Anglian and south-coast marshes. On the sandier marshes of the west and north-east coasts, however, it is often dominant over wide areas and sometimes replaces *Salicornia* as pioneer colonist. Its abundance is less closely related to its frequency of immersion than is the case with many halophytes, and it occurs to a certain extent at all levels in some salt-marshes. It may also sometimes be found on rocky cliff-ledges, particularly in Scotland. The large areas of *Puccinellietum* found along the west coast provide good grazing.

It grows in thick tufts from a creeping perennial rootstock, producing numerous trailing, and sometimes rooting, stems. The somewhat fleshy leaves are rather like those of a rush, the margins being inrolled. The round, smooth flowering-stems are a foot or more in height, the flowers being arranged in a narrow, rather stiff panicle. Flowers: July-August.

SAND-FESCUE, *Festuca rubra* var: *arenaria.* This well-known variety of red fescue (sometimes elevated to the rank of a species as *Festuca arenaria*) is a common inhabitant of most sand-dunes. It is usually the first grass to appear in the shifting sand of open marram dunes and, with its common associate, the sand-sedge, is a most efficient agent in stabilising the surface. At a later phase, when the surface has become almost completely covered by vegetation and the marram-grass is being choked out, it may replace it as the dominant species. It is also frequently instrumental, again with the help of the sand-sedge, in recolonising " blow-outs," particularly when they occur some distance from an area of actively growing marram-grass.

It is a perennial, with extremely extensive creeping roots, which make it an excellent sand-binder. It is rather coarser in growth than other varieties, and produces flowering stems up to 2 feet in height. The leaves are up to 6 in. long, thick and rigid, with inrolled margins. The flowers occur in a rather large spreading panicle, usually slightly drooping. Flowers: June-July.

Apart from this sand-dune variety, red fescue is often abundant in other coastal habitats, as it is in inland ones. In particular, it often dominates a zone on the higher levels of salt-marshes, especially where the soil is sandy. Thus in the West it often replaces the sea manna-grass as dominant, being associated with such species as sea-milkwort, sea-plantain, and one or other of the maritime rushes. It is also frequently the commonest grass on cliff-tops and, in exposed and ungrazed situations, may form largely pure communities.

SEA COUCH-GRASS, *Agropyron junceiforme* (*junceum*). The most characteristic habitat of this well-known grass is the mobile sand above the high-tide mark on sandy beaches. It is an efficient sand-binder, and can tolerate a certain amount of exposure to sea-water, unlike marram-grass. Thus it often forms foredunes in front of the main marram dunes, but these are comparatively low because it does not possess the same power of growing upwards through loose sand. It is commonly distributed round the whole coastline, although in a number of dune areas marram-grass is the sole dune-building agent. In one or two small areas of blown sand, however, particularly in Ireland, it is entirely responsible for the existence of dunes, marram-grass being completely absent.

It is a perennial, resembling in many ways the common couch-grass or twitch (*Agropyron repens*). It has extensive creeping roots, and produces many prostrate barren shoots and erect flowering stems, about 2 feet in height. These are smooth, rigid, and solid in their upper part. The leaves are rather thick and stiff, with the margins inrolled as in other sand xerophytes, and end in a sharp point. Their inside surface shows prominent ribs, which are covered with short, white hairs. Both stems and leaves are glaucous. The flowers occur in a stout spike, 2-5 in. long, with rather distant spikelets containing 4-5 flowers. Flowers: July-August.

SHARP COUCH-GRASS, *Agropyron pungens*. This species is very similar to the last, and may be found near the upper edges of salt-marshes,

John Markham

Plate 16   Sea-purslane, *Halimione portulacoides;* a common salt-marsh plant

along the banks of tidal rivers, and sometimes on stable dunes or shingle. It is widely distributed, but is absent from long stretches of the coastline.

It differs from *Agropyron junceiforme* chiefly in the rather broader leaves, which have very thick, rough, but not hairy, ribs. The rather larger spikelets occur closer together on the spike, and contain more flowers. Flowers: July-August.

There is also a common hybrid between the two species, usually referred to as *Agropyron laxum*, and showing intermediate characteristics.

SEA HARD-GRASS, *Parapholis strigosa* (formerly *Lepturus filiformis*). This little grass is frequently seen on the higher levels of salt-marshes (e.g. in the red fescue zone in the West), on stable shingle banks, or along the sides of tidal rivers. It is abundant all round the Mediterranean, frequent in the southern half of England, but becomes less common northwards.

In general appearance it is rather like a small edition of the common rye-grass (*Lolium perenne*), but it is an annual with a fibrous root. The stems are much-branched, semi-prostrate at first, and then curving upwards to a height of 4-12 in. The leaves are short and narrow, rather leathery and rough, and usually somewhat inrolled. The flowers occur in a stiff, very slender spike, which is often curved. The anthers are yellowish-white. Flowers: July-September.

Parapholis incurva is a very similar species, with a stouter stem and spike, the latter being more definitely curved, found chiefly in the South-West. Flowers rather earlier.

SEA LYME-GRASS, *Elymus arenarius*. This very distinct species grows in similar positions to the sea couch-grass on sandy beaches, although it is not so common. Like this grass, it can endure exposure to sea-water, and sometimes forms low foredunes on the seaward side of the main marram-grass dunes. It is an efficient sand-binder, and has been planted for this purpose in some places, notably in Holland. It occasionally invades shingle beaches, and I have also seen it growing on rocky cliffs in Scotland. A common plant of maritime sands in the temperate and colder regions of the northern hemisphere, it is thinly distributed round much of the British coast, becoming rarer in the South.

It generally grows in large round tufts from an extensive, perennial,

F.C.   P creeping rootstock. Like other dune grasses, it has vigorous powers of producing new shoots when buried. The stout flowering stems are smooth and round, 2-4 feet in height. The leaves are extremely stiff and rather broad; the upper surface being bluish-green and showing prominent ribs, the under surface smooth and light-green. The flowers occur in a dense spike, 6 in. or more long, the spikelets often being purplish. Flowers: July-August.

SEA-BARLEY or SQUIRREL-TAIL GRASS, *Hordeum marinum*. This small annual grass is not uncommon along parts of the east coast, but is rarer on the south coast and absent from Scotland and most of the West. It is widely distributed in western Europe and round the Mediterranean. It appears characteristically along the upper edges of salt-marshes, sometimes spreading into ordinary grazing land near the sea, notably along the Wash, and is occasionally found on mature sand-dunes.

Very similar to the common wall-barley (*Hordeum murinum*), it is a smaller plant and distinctly glaucous. The smooth, round stems are often semi-prostrate, and bear numerous short, flat, rather rough leaves. In favourable situations the stems are about a foot long, but are usually less. The thick flower-spikes are smaller and the awns a little shorter than those of the common species. Flowers: June-July.

OTHER SUBMARITIME GRASSES

A considerable number of other local grasses are mainly confined to the coastal belt in this country. From these I have selected two as being typical:

Desmazeria marina (formerly *Festuca rottboellioides*) is common in western Europe and the Mediterranean region. It is scattered round the coast of Britain on sand-dunes, and sometimes on shingle or rocky cliff-ledges. It is a small, rather stiff annual, growing in tufts and producing erect or slightly curved stems, 4-8 in. high. The small leaves are narrow and flat, with prominent ribs. The spike-like panicle is rather dense, the almost sessile spikelets being attached alternately as in a small couch-grass spike, but all turning one way. Flowers: May-July.

Vulpia membranacea (formerly *Festuca uniglumis*) is also common in western Europe and the Mediterranean. It is probably most frequent in the South-West, but is scattered round our western, southern and

south-eastern coasts on sand-dunes. It is an annual, seldom above
6 in. high, forming little tufts. The slender stems are stiff and mainly
erect, but are sometimes prostrate at the base. The lower leaves are
narrow and bristle-like, with inrolled margins, but the upper ones are
broader. The cylindrical panicle is short, dense and usually one-sided.
The long awns are a prominent feature. Flowers: June-July.

Other interesting submaritime species are *Alopecurus bulbosus,*
Cynodon dactylon, Gastridium ventricosum (lendigerum), Polypogon mon-
speliensis and *Poa bulbosa.*

POLYPODIACEAE

Sea-spleenwort, *Asplenium marinum* (Pl. XL, p. 207). This fern is
usually found in a shady crevice on a rocky cliff or sea-wall, or near
the mouth of a cave, growing from the sides or the roof. A west
European species, it is the only British maritime fern and is reasonably
common down the whole west coast (including Scotland) and along
the south coast as far as Hampshire. On the east coast it is not found
farther south than Yorkshire.

Its long woody rootstock, clothed with purplish scales, is generally
firmly wedged into a crack in the rocks, into which its wiry roots
penetrate deeply. It differs from other spleenworts in its large and
leathery fronds, which are commonly about 6 in. long, but in luxuriant
specimens may measure over a foot. These are borne on rather thick,
polished stalks, and the roughly oval pinnae, an inch or more long,
are slightly toothed. The sori are large, rust-coloured, and covered
with a thick meal.

GLOSSARY

adventitious roots: those which do not proceed from the base of the main stem or rootstock.

aerial stems: stems which are above the surface of the ground.

agg.: this abbreviation after a species-name indicates that the name refers to an aggregate species, which has later been divided into several species, sub-species or other categories.

Agropyretum: a plant community dominated by *Agropyron junceiforme* (sea couch-grass).

alien: a plant introduced from abroad which is not a native species.

alternate: arranged alternately on either side of the stem, not in opposite pairs.

Ammophiletum: a plant community dominated by *Ammophila arenaria* (marram-grass).

anther: the upper part of a stamen, where the pollen is produced.

aqueous tissue: the parts of some plants which are composed of water-holding cells, not containing chlorophyll.

Armerietum: a plant community dominated by *Armeria maritima* (thrift).

Asteretum: a plant community dominated by *Aster tripolium* (sea-aster).

awn: a bristle-like point, attached to the glumes of some grasses.

axil: the angle between a leaf and the stem, above the former's point of attachment.

beak: a pointed projection.

biennial: a plant normally living for two years, which produces leaves and perhaps a fleshy root during the first year, but flowers only in the second year and then dies.

biotic: concerned with living organisms.

blow-out: an area of open sand amongst partly fixed dunes caused by the wind removing the surface-cover of vegetation.

bract: a modified leaf, usually smaller and of a different shape from the foliage leaves, which is closely associated with a flower or inflorescence.

calcicole: characteristic of soils with a high lime-content.

calcifuge: characteristic of acidic or non-limy soils.

calyx: the sepals considered as a whole, whether free or joined together.

capsule: a dry fruit which splits open when ripe, formed from two or more carpels.

carbon assimilation: see photosynthesis.

carpel: the whole female organ, consisting of a lower portion, the *ovary*, and an upper portion upon which the pollen falls, the *stigma*, usually roughened or sticky. These two are generally connected by a short stalk, the *style*.

carr: woodland or scrub developed on swampy land, seen typically in the fenland.

casual: any species which appears occasionally in a plant community but is not really characteristic.

chlorophyll: the green colouring matter present in most plants.

cladode: a modified branch, containing chlorophyll, and functioning like a leaf.

climatic climax: the ultimate type of vegetation produced by succession in a particular climate; also called the climax formation.

closed community: a plant community where the vegetation is more or less continuous.

community: see plant community.

consociation: see plant consociation.

constant species: a species, other than a dominant, which is nearly always present in a plant community.

corolla: the petals considered as a whole, whether free or united.

cuticle: the outer layer on the walls of many cells, particularly those on leaf-surfaces, which makes them impervious to water and often to gasses also.

disc: the inner group of tubular florets in the compact inflorescence of a composite " flower."

dominant species: the species mainly responsible for the general appearance of a plant community, usually the tallest or commonest plant present.

edaphic: concerned with the soil.

entire: without teeth or lobes.

female flower: a flower having a pistil, but no stamens.

Festucetum rubrae: a plant community dominated by *Festuca rubra* (red fescue).

floret: an individual flower in a crowded or compound inflorescence.

formation: see plant formation.

frond: the " leaf " of a fern or seaweed.

glabrous: without hairs.

glaucous: pale bluish-green, sometimes with a waxy bloom.

glume: one of the chaffy bracts which enclose the flowers of sedges and grasses.

habitat: the place where a plant grows, plus all those conditions connected with it which affect the plant's growth.

Halimionetum: a plant community dominated by *Halimione* (*Obione*) *portulacoides* (sea-purslane); formerly known as *Obionetum*.

halophyte: a plant which can live where the water is salt. This class is sometimes divided into " true halophytes " which thrive in places where the soil-water is permanently saline, and " spray halophytes " which grow in habitats exposed to salt spray, but are not normally submerged by sea-water.

halosere: the series of communities which replace each other in a salt-marsh.

humus: the dark organic matter in the soil formed by the decay of plant or animal remains.

inflorescence: a more or less compact group of flowers considered collectively.

Juncetum gerardii: a plant community dominated by *Juncus gerardi* (mud-rush).

Juncetum maritimae: a plant community dominated by *Juncus maritimus* (sea-rush), often referred to simply as *Juncetum*.

lanceolate: somewhat tapered at both ends (lance-shaped).

leaflet: one of the divisions of a compound leaf.

Limonietum: a plant community dominated by *Limonium vulgare* (sea-lavender).

linear: very narrow.

lobe: one of the irregular portions of a deeply indented leaf or petal.

male flower: a flower having stamens, but with the pistil absent or not fully developed.

mesophyte: a normal plant, living under conditions which are neither very wet nor very dry.

node: a point on a stem, often thickened, to which a leaf or leaves are attached.

open community: a plant community where the vegetation is not continuous.

opposite: arising in pairs, one on either side of the stem.

osmotic pressure: the suction exerted by the root-hairs, which enables a plant to absorb water and dissolved salts from the soil.

panicle: a branched raceme; i.e. an inflorescence in which the branches (themselves racemes) are arranged like the flowers in a simple raceme.

perennial: a plant normally living for at least three years.

perianth: the calyx and corolla, considered as one unit.

photosynthesis: the process by which the green cells build up starch and other carbon compounds, under the influence of light, by absorbing carbon dioxide.

Phragmitetum: a plant community dominated by *Phragmites communis* (common reed).

pinna (pl. *pinnae*): the primary division of a compound leaf or frond.

pinnate: describes a compound leaf in which the pinnae or leaflets are arranged in two ranks on opposite sides of the stem.

pinnatifid: lobed, with the lobes arranged in a pinnate manner.

pioneer: a first colonist of a bare habitat.

pistil: the carpel or carpels of a flower, whether free or united.

plant community: any collection of plants growing together which has a recognisable individuality.

plant consociation: a plant community having (usually) a single dominant species.

plant formation: the main type of vegetation produced in a habitat under particular climatic conditions. It is the largest unit of vegetation recognised, and the same dominant life-forms are common to a formation wherever it may occur in the world.

plant society: a local plant community with a separate dominant species found within a consociation, usually owing its origin to some small difference in habitat.

Plantaginetum: a plant community dominated by *Plantago maritima* (sea-plantain); more correctly *Plantaginetum maritimae.*

psammosere: the series of communities which replace each other on blown sand.

Puccinellietum: a plant community dominated by *Puccinellia (Glyceria) maritima* (sea manna-grass); formerly known as *Glycerietum.*

raceme: an inflorescence in which a number of stalked flowers are arranged along an unbranched stem which continues to grow during the development of the flowers, the oldest flowers being at the base and the youngest at the tip.

radical: attached to the stem at or below ground-level.

ray: the outer florets in the compact inflorescence of a composite " flower."

rhizoids: the root-like hairs by which mosses and liverworts are attached to the ground.

rhizome: an underground or buried stem, usually root-like in appearance.

rootstock: the portion of a perennial's stem, more or less underground, which persists from one year to another.

runner: a branch from a stem which creeps along the surface of the ground, usually taking root at some distance from the parent plant.

Salicetum repentis: a plant community dominated by *Salix repens* (creeping willow).

Salicornietum: a plant community dominated by *Salicornia stricta* (*herbacea*) (glasswort) or by some related annual species of *Salicornia*.

salt-pan: a depression in a salt-marsh, not connected to a drainage channel, which is periodically flooded with sea-water.

Scirpetum maritimae: a plant community dominated by *Scirpus maritimus* (sea club-rush).

sepals: the individual floral leaves which together form the calyx.

sere: any natural series of communities which replace each other in a particular habitat.

sessile: without stalks.

society: see plant society.

sorus (pl. *sori*): a small heap of minute capsules (*sporangia*), which contain the spores, situated on the back of a fern frond, and often covered with an outgrowth from the cuticle known as the *indusium*.

Spartinetum: a plant community dominated by *Spartina townsendii* (rice-grass).

spike: an inflorescence resembling a raceme, except that the flowers are sessile and are directly joined to the stem.

spikelet: one of the small, compact spikes in which the glumes and flowers of sedges and grasses are arranged.

spur: a hollow projection on a petal or sepal, resembling a cock's spur.

stamen: the male organ which produces the pollen, usually consisting of a *filament* or stalk bearing an *anther* at its top.

stigma (pl. *stigmata*): see carpel.

stolon: a modified stem, similar to a rhizome, but usually running along the surface or just beneath it.

stoma (pl. *stomata*): a pore on the surface of a leaf or stem through which gases can be exchanged.

style: see carpel.

submaritime species: plants, other than halophytes or dune-species, rarely found more than a few miles from the sea.

succession: the natural replacement of one community by another, as the nature of a habitat alters.

taproot: a main, thickened root, descending perpendicularly into the ground without any but quite small fibrous branches.

transpiration: the process by which water-vapour is evaporated from the leaves.

transpiration-checks: devices, usually on leaves, whereby the rate of transpiration is reduced.

umbel : an inflorescence in which the stalks of all the flowers radiate from a single point. In a " compound umbel " the stalks of a number of simple or " partial " umbels also meet at a point.

whorl: a group of leaves or flowers arising from the same point on a stem in the form of a ring.

wing: the continuation of the edge of a leaf down a stem to form a raised border, as in certain thistles.

winter annual: a plant which germinates in the autumn, flowering and setting seed the following year before the summer has reached its height.

xerophyte: a plant adapted to grow under dry conditions.

Zosteretum: a plant community dominated by *Zostera* species (eel-grass).

BIBLIOGRAPHY

IN THE COURSE of writing this book I have consulted a large number of original papers and books, and should like to take this opportunity of making grateful acknowledgment to the various authors whose names appear in the following list. For the benefit of those wishing to follow up particular points in more detail, I have arranged the literature as far as possible under the main habitats with which it is concerned. As a number of papers deal with several different habitats at the same time (e.g. those describing the vegetation along the East Anglian coast), this classification is bound to be rather arbitrary. I have, however, endeavoured to place those papers concerned with more than one habitat under the heading covering their principal contents (usually with a mention of the other habitats involved). I have included in the list some more general papers dealing with relevant topics and have also added a list of books which can be recommended for the identification of species.

GENERAL

CAREY, A. E., and OLIVER, F. W. (1918). Tidal lands: a study of shore problems. London, Blackie and Sons.

CHAPMAN, V. J. (1936). The halophyte problem. *Quart. Rev. Biol.*, 11: 209. (1942) A new perspective on the halophytes. *Quart. Rev. Biol.*, 17: 291.

DELF, E. M. (1911). Transpiration and the behaviour of stomata in halophytes. *Ann. Bot.*, 25: 485.

(1912). Transpiration in succulent plants. *Ann. Bot.*, 26: 409.

(1915). The meaning of xerophily. *J. Ecol.*, 3: 110.

FRITSCH, F. E., and SALISBURY, E. J. (1938). Plant form and function. London, G. Bell and Son.

HILL, T. G. (1908). On the osmotic properties of the root-hairs of certain salt-marsh plants. *New Phytol.*, 7: 133.

MACDOUGAL, D. T., RICHARDS, H. M., and SPOEHR, H. A. (1919). The basis of succulence in plants. *Bot. Gaz.*, 68: 22.

MAXIMOV, N. A. (1929). The plant in relation to water. Edited by R. H. Yapp. London. Allen and Unwin.

(1931). Symposium on Xerophily. *J. Ecol.*, 19: 273.

PRAEGER, R. L. (1934). The Botanist in Ireland. Dublin, Hodges, Figgis and Co.

STEERS, J. A. (edited by) (1934). Scolt Head Island. Cambridge, W. Heffer and Sons.

(1946). The coastline of England and Wales. Cambridge, University Press.

(*This book contains full references to the principal papers on the physiography of the coast.*)

TANSLEY, A. G. (1923). Practical Plant Ecology. London, Allen and Unwin. (Republished 1946 under the title: Introduction to Plant Ecology.)

(1939). The British Islands and their Vegetation. Cambridge, University Press. (*The classic book on plant ecology*).

(1949). Britain's Green Mantle. London, Allen and Unwin.

SALT-MARSHES

CARTER, N. (1932). A comparative study of the alga flora of two salt-marshes. *J. Ecol.*, 20: 341, 21: 128 and 385.

CHAPMAN, V. J. (1938-41). Studies in salt-marsh ecology. Parts I-VIII. *J. Ecol.*, 26: 144, 27: 160, 28: 118, 29: 69. (*The most important modern papers on salt-marsh ecology.*)

(1938). Marsh development in Norfolk. *Trans. Nfk. Norw. Nat. Soc.*, 14: 394.

CONWAY, V. (1933). Further observations on the salt-marsh at Holme-next-the-sea. *J. Ecol.*, 21: 263.

HESLOP HARRISON, J. W. (1918). A survey of the lower Tees marshes and of the reclaimed areas adjoining them. *Trans. Nat. Hist. Soc. Northumb.*, 5: 89.

MARSH, A. S. (1915). The maritime ecology of Holme-next-the-sea, Norfolk (also sand-dunes and shingle), *J. Ecol.*, 3: 65.

McCREA, R. H. (1926). The salt-marsh vegetation of Little Island, Co. Cork. *J. Ecol.*, 14: 342.

MORSS, W. L. (1927). The plant colonisation of the Merse Lands in the estuary of the River Nith. *J. Ecol.*, 15: 310.

NEWMAN, L. F., and WALWORTH, G. (1919). A preliminary note on the ecology of part of the South Lincolnshire coast. *J. Ecol.*, 7: 204.

OLIVER, F. W. (1920). *Spartina* problems. *Ann. Appl. Biol.*, 7: 57.

(1925). *Spartina townsendii;* its mode of establishment, economic uses and taxonomic status. *J. Ecol.*, 13: 74.

PEACE, T. R. (1928). Further changes in the salt-marsh and sand-dunes of Holme-next-the-sea (also shingle), *J. Ecol.*, 16: 412.

PHILIP, G. (1936). An enalid plant association in the Humber Estuary. *J. Ecol.*, 24: 205.

PRIESTLEY, J. H. (1911). The pelophilous formation of the left bank of the Severn Estuary. *Bristol Nat. Proc.*, 4th series, 3: part 1.

RICHARDS, F. J. (1934). The salt-marshes of the Dovey Estuary; the rates of vertical accretion, horizontal extension and scarp erosion. *Ann. Bot.*, 48: 89.

ROWAN, W. (1913). Note on the food-plants of rabbits on Blakeney Point, Norfolk (also shingle and dunes), *J. Ecol.*, 1: 273.

TANSLEY, A. G. (1941). Note on the status of salt-marsh vegetation. *J. Ecol.*, 29: 212.

WADHAM, S. M. (1920). Changes in the salt-marsh and sand-dunes of Holme-next-the-sea (also shingle), *J. Ecol.*, 8: 232.

WEIHE, P. O. (1935). A quantitative study of the influence of tide upon populations of *Salicornia europea* (*stricta*). *J. Ecol.*, 23: 323.

YAPP, R. H., JOHNS, D., and JONES, O. T. (1916). The salt-marshes of the Dovey Estuary. Parts I and II. *J. Ecol.*, 4: 27, and 5: 65.

YAPP, R. H. (1922). The Dovey salt-marshes in 1921. *J. Ecol.*, 10: 18.

SAND-DUNES

BARRETT, W. H. (1940). The composition and properties of shore and dune soils. *Geol. Mag.*, 77: 383.

BRUCE, E. M. (1931). The vegetation of the sand-dunes between Embleton and Newton. *Vasculum*, 17: 94.

ELLISTON-WRIGHT, F. R. (1932). Alterations in vegetative growth due to environmental adaptation in Braunton Burrows. *Bot. Exch. Club 1932 Report*: 258.

GIMINGHAM, C. H., GEMMELL, A. R., and GREIG-SMITH, P. (1948). The vegetation of a sand-dune system in the Outer Hebrides. *Trans. Bot. Soc. Edinb.*, 35: 82.

GOOD, R. (1935). The ecology of the flowering plants of the South Haven peninsula, Studland Heath, Dorset. *J. Ecol.*, 23: 361.

HARTLEY, J. W., and WELDON, J. A. (1914). The Manx sand-dune flora. *J. Bot.*, 52: 170.

HEPBURN, I. (1945). The vegetation of the sand-dunes of the Camel estuary, North Cornwall. *J. Ecol.*, 32: 180.

HILL, J. E. (1927). Holy Island. *Vasculum*, 14: 94.
　　　　　　　Ross Links. *Vasculum*, 14: 121.

McLEOD, A. M. (1948). Some aspects of the plant ecology of the Island of Barra. *Trans. Bot. Soc. Edinb.*, 35: 48.

MOORE, E. J. (1931). The ecology of the Ayreland of Bride, Isle of Man (also shingle), *J. Ecol.*, 19: 115.

MOSS, C. E. (1907). The geographical distribution of vegetation in Somerset: Bath and Bridgewater district (also salt-marshes), *Royal Geog. Soc. Supplement*, 1907.

ORR, M. Y. (1912). Kenfig Burrows. Scot. Bot. Rev., 1: 209.

OVINGTON, J. D. (1950). The afforestation of the Culbin Sands. *J. Ecol.*, 38: 303.

PATTON, D., and STEWART, E. J. A. (1914). The Flora of the Culbin Sands. *Trans. Bot. Soc. Edinb.*, 26: 345.

—— (1924). Additional notes on the flora of the Culbin Sands. *Trans. Bot. Soc. Edinb.*, 29: 27.

PEARSALL, W. H. (1934). North Lancashire sand-dunes. *Naturalist*, 1934: 201.

RICHARDS, P. W. (1929). Notes on the ecology of the bryophytes and lichens at Blakeney Point, Norfolk (also shingle), *J. Ecol.*, 17: 127.

SALISBURY, E. J. (1920). The significance of the calcicolous habit. *J. Ecol.*, 8: 202.

—— (1922). The soils of Blakeney Point: a study of soil reaction and succession in relation to the plant-covering (also shingle), *Ann. Bot.*, 10: 391.

—— (1925). Note on the edaphic succession in some dune soils with special reference to the time factor. *J. Ecol.*, 13: 322.

SKINNER, E. (1934). A survey of the dunes between Meggie's Burn and Seaton Sluice. *Vasculum*, 20: 122.

SMITH, W. G. (1905). The flora of the Forfar and Fife sand-dunes. *Scot. Geog. Mag.*, 21: 70.

STARR, A. M. (1912). The comparative anatomy of dune plants. *Bot. Gaz.*, 54: 265.

STEERS, J. A. (1937). The Culbin Sands and Burghead Bay. *Geog. Journ.*, 90: 498.

THOMPSON, H. S. (1922). Changes in the coast vegetation near Berrow, Somerset (also salt-marshes), *J. Ecol.*, 10: 53.

—— (1930). Further changes in the coast vegetation near Berrow, Somerset. *J. Ecol.*, 18: 126.

TRAVIS, W. G. (1916). The flora of the Lancashire dunes. *Lancs and Cheshire Nat.*, 9: 29, 83.

WATSON, W. (1918). Cryptogamic vegetation of the sand-dunes of the west coast of England. *J. Ecol.*, 6: 126.

Note: *Downs and Dunes* by E. J. Salisbury (London, G. Bell and Sons, 1952.) has unfortunately not yet appeared as we go to press, but it is certain to be a most stimulating book.

SHINGLE BEACHES

HILL, T. G., and HANLEY, J. A. (1914). The structure and water-content of shingle beaches. *J. Ecol.*, 2: 21.

McLEAN, R. C. (1915). The ecology of the maritime lichens at Blakeney Point, Norfolk (also sand-dunes), *J. Ecol.*, 3: 129.

OLIVER, F. W. (1912). The shingle beach as a plant habitat. *New Phytol.*, 11: 73.

(1913). Some remarks on Blakeney Point, Norfolk. *J. Ecol.*, 1: 4 (also salt-marshes and sand-dunes).

(1913-1929). Blakeney Point Reports (also salt-marshes and sand-dunes). *Trans. Nfk. and Norw. Nat. Soc.*

OLIVER, F. W., and SALISBURY, E. J. (1913). Vegetation and mobile ground as illustrated by *Suaeda fruticosa* on shingle. *J. Ecol.*, 1: 249.

(1913). Topography and vegetation of Blakeney Point (also salt-marshes and dunes). *Trans. Nfk. and Norw. Nat. Soc.*, 9 (also issued separately).

WATSON, W. 1922. List of lichens, etc., from Chesil Beach. *J. Ecol.*, 10: 255.

CLIFFS AND CLIFF-GRASSLAND

ASPREY, G. F. (1947). The vegetation of the Islands of Canva and Sanday, Inverness-shire. *J. Ecol.*, 34: 182.

BOWEN, E. J. (1930). A survey of the flora of the North Gower coast. *Proc. Swansea Sci. and Field Nat. Soc.*, 1 1930: 109.

CRAMPTON, C. B. (1911). The vegetation of Caithness considered in relation to the geology. British Vegetation Committee, 1911. (Obtainable from Cambridge University Press.)

GIMINGHAM, C.H., and ROBERTSON, E. T. (1951). Contributions to the Maritime Ecology of St. Cyrus, Kincardineshire. *Trans. Bot. Soc. Edinb.*, 35: 370.

HEPBURN, I. (1943). A study of the vegetation of sea-cliffs in north Cornwall. *J. Ecol.*, 31: 30.

HILL, J. E. (1927). Among the Farnes. *The Vasculum*, 14: 1.

McLEAN, R. C. (1935). An ungrazed grassland on limestone in Wales. *J. Ecol.*, 23: 436.

PETCH, C. P. (1933). The Vegetation of St. Kilda. *J. Ecol.*, 21: 92.

POORE, M. E. D., and ROBERTSON, V. C. (1949). The Vegetation of St. Kilda in 1948. *J. Ecol.*, 37: 82.

TANSLEY, A. G., and ADAMSON, R. S. (1926). A preliminary survey of the chalk grasslands of the Sussex downs. *J. Ecol.*, 14: 10.

TURRILL, W. B (1927). The flora of St. Kilda. *Bot. Exch. Club 1927 Report:* 428.
VEVERS, H. G. (1936). The land vegetation of Ailsa Craig. *J. Ecol.,* 24: 424.
WILMOTT, A. J., and CAMPBELL, M. S. (1945). The flora of Uig. Arbroath, T. Buncle and Co.

FLORAS

VASCULAR PLANTS:

BENTHAM, G., and HOOKER, J. D. (1924). Handbook of the British Flora (7th Edition, revised by A. B. Rendle). London, L. Reeve and Co.
BUTCHER, R. W., and STRUDWICK, F. E. (1944). Further Illustrations of British Plants (2nd edition). Ashford, L. Reeve and Co.
CLAPHAM, A. R., TUTIN, T. G., and WARBURG, E. F. (1952). Flora of the British Isles. Cambridge, University Press.

(*For many years there has been a great need for an up-to-date flora of the British Isles, and this fine book fills the gap admirably.*)
Of the older books, the following can be recommended:
COSTE, H. (1901-1906). Flore de la France (reissued, 1938), Paris.
(*This excellent book contains practically all the British plants, and many European species likely to be found in the British Isles as aliens.*)
DRUCE, G. C. (1930). Hayward's Botanist's Pocket-book, 19th edition. London, Bell and Sons.
(*Useful in the field.*)
FITCH, W. H., and SMITH, W. G. (1924). Illustrations of the British Flora (7th edition). London, L. Reeve and Co.
HOOKER, J. D. (1930). The Student's Flora of the British Isles (3rd edition, reprinted). London, Macmillan and Co.

In addition to the above works on vascular plants, the following are in process of publication in separate parts:
CLAPHAM, A. R., PEARSALL, W. H., and RICHARDS, P. W. (edited by). Biological Flora of the British Isles. Reprinted from the *Journal of Ecology,* and available in separate parts, each dealing with one species. Cambridge University Press.
ROSS-CRAIG, S. Drawings of British Plants. London, Bell and Sons.

MOSSES:

Dixon, H. N., 1924. The Student's Handbook of British Mosses (3rd edition). Eastbourne, V. V. Sumfield.

LICHENS:

Lorrain Smith, A. (1921). A Handbook of British Lichens. London, British Museum.

ALGAE:

Newton, L. (1931). A Handbook of British Seaweeds. London, British Museum.

INDEX

Figures in heavy type refer to pages opposite which illustrations appear. Figures in brackets refer to pages where a plant is referred to under an English but not under a Latin name or vice versa. Figures in italics indicate the descriptions of species included in Chapter 12. Place names are indexed only when of special importance.

N.B. All plants whose English name begins with "sea" are listed together.

THE NEW NATURALIST
A Survey of British Natural History

See next page Special Volumes and Monographs published in the New Naturalist library

Special Volumes

LORDS AND LADIES—*Cecil T. Prime*

" A triumphant success. The book is packed with curious information about this eccentric plant." THE TIMES EDUCATIONAL SUPPLEMENT

" A joy to the naturalists, but much of it can almost be described as light reading. The private life of lords-and-ladies fully justifies Dr. Prime's amused and thorough treatment." *Eric Keown,* PUNCH

" Not being a botanist, I should not have believed it possible for me to read with enjoyment a book entirely devoted to one plant and its close allies. But Dr. Prime is no ordinary botanist. He has produced an exceptionally readable book." THE FIELD

WILD ORCHIDS OF BRITAIN (with a key to the species)—*V. S. Summerhayes*

" Almost everything that anyone can want to know about the British orchids is here. An excellent key is provided. This is a book that demands several readings. It is destined to be the standard work on the subject. But it is much more interesting and much more vital than those awful words ' standard work ' would lead anyone to suppose. I, at least, was fascinated."

Brian Vesey-FitzGerald, OUT OF DOORS

" Really the first serious monograph on the native orchids to be published at a reasonable price. It should be (and without any doubt will be) in the hands of all British orchidomaniacs. Mr. Summerhayes has written an extremely scholarly and comprehensive monograph."   *Jocelyn Brooke*, TIME AND TIDE

" A botanical work of outstanding importance. It will appeal to all wild-flower-loving countrymen."   COUNTRYMAN

WILD FLOWERS—*John Gilmour & Max Walters*

" This authoritative book, by the director of the Cambridge University Botanic Garden and the Curator of the Cambridge University Herbarium, provides a mass of fascinating information. It is something far more than the usual descriptive reference book, for it includes much historical, biological, ecological and geological learning, lucidly presented to the amateur reader.

This is a truly important book, and a serious one written by serious botanists; at the same time it is a book to be read with pleasure by any amateur botanist or lover of the wild flowers of our country."

V. Sackville-West, THE OBSERVER

" We have here a flower-book most thoroughly to be recommended to all botanists and to every naturalist with any interest in the wild plants of these islands."

BIRMINGHAM POST

MOUNTAIN FLOWERS—*John Raven & Max Walters*

" All who love mountain flowers will be delighted by this book . . . It will please and further instruct the initiated and give a lifelong interest to many lovers of our mountains who have not yet learnt to know their delicately beautiful plants."

THE TIMES LITERARY SUPPLEMENT

" For those who wish to sample the delights of our own mountains, Mr. Raven and Dr. Walters have written a first-class companion . . . I have no hesitation in advising all mountain-climbing flower lovers and flower-loving mountaineers to put it in their rucksacks."

R. S. R. Fitter, BIRMINGHAM POST

WILD FLOWERS OF CHALK AND LIMESTONE
—J. E. Lousley

" The sixteenth volume in the New Naturalist Library attains the high standard its predecessors have led us to expect . . . Mr. Lousley takes us on an enjoyable botanical tour of the principal chalk and limestone regions . . . well informed and enthusiastic guide." GLASGOW HERALD

" One is tempted, as each new volume in the New Naturalist appears, to hail it as the best yet published (which is compliment enough to the series as a whole), but even when one remembers that that temptation exists, it would be almost impossible to praise this latest edition too highly . . . This is a wonderfully friendly book, a book to keep and treasure, to read at a sitting and then to dip into again and again. The illustrations are quite beyond praise." THE FIELD

TREES, WOODS AND MAN—H. L. Edlin

" With this handsome addition to their New Naturalist series, the publishers have once again succeeded in producing a book of wide general interest by an authoritative writer who neither condescends to the reader nor blinds him with science." THE OBSERVER

" This excellent and finely illustrated book should be used by everyone concerned for the future of this country. The book is technical enough to satisfy both the expert and the amateur of forestry." GLASGOW HERALD

THE WORLD OF THE SOIL—Sir E. John Russell, F.R.S.

". . . An intensly interesting and very knowledgeable book . . . his wide knowledge of the dozen sciences involved in the study of soil continually amazes the reader . . . The reasons for cultivating, manuring, draining, etc., are explained with great lucidity and a remarkable economy of words; and the interest is maintained by a skilful intermingling of the history of agriculture and science . . . one of the New Naturalist Library, maintains the high standard set by earlier issues of the series. The illustrations, in particular, are striking and excellently chosen." THE TIMES LITERARY SUPPLEMENT

THE ART OF BOTANICAL ILLUSTRATION
—Wilfred Blunt

" The text is as interesting and lively as the illustrations, and one compact mass of information . . . To many, indeed to most, readers it is a new world of beauty . . .

I hope that I am conveying something of the excitement of reading and looking through this book of flower drawings . . . unlikely to be superseded, except for particular studies, for a very long time." *Sacheverell Sitwell*, SPECTATOR

" Mr. Wilfred Blunt's book is unique . . . a classic—it will be read and admired to-day, and will be kept for reference for all time."

Lord Aberconway, THE SUNDAY TIMES

" This beautiful book is unique of its kind . . . Let no one think this is a book for the specialist. It is essential for the specialist, certainly, but it is for all flower-lovers." *V. Sackville-West*, THE OBSERVER

BRITISH PLANT LIFE—*W. B. Turrill*

" A fascinating and highly successful attempt to explain to the intelligent reader some of the factors that have determined the range and form of British plants, and the means by which we can learn more of these factors . . . Schools and colleges will find this book indispensable . . . It is indeed a triumph that one volume should in text and illustrations unite so much beauty with the strict discipline of science." WESTERN MAIL

" We have now come to look forward to new volumes in the series, well written, beautifully produced and awakening interest in some fresh feature of Britain's countryside . . . In every way up to the standards of its predecessors . . . a clear and substantial exposition of plant geography in Britain . . . deserves a wide and varied public." MANCHESTER GUARDIAN

THE SEA SHORE—*C. M. Yonge*

" By far the best book that has yet appeared in this series of good books. Here is one that you simply cannot put down, beautifully illustrated, beautifully and simply written, crammed full of excitement and oddity and wonder. Everyone who has ever been to the sea shore should read it and learn what they missed, and it is a ' must ' for all those who hope to have a holiday by the sea next year." THE FIELD

" The most enchanting yet of the delightful New Naturalist series . . . It is a book to take on a seaside holiday to enrich the interest in familiar plants and animals . . . These pictures leave one breathless with delight." BIRMINGHAM POST

" A philosopher could brood for years over these pages confronted with the mystery of a life-force seeking to establish itself in countless ways. A child could gaze enraptured at the magic life revealed in the pictures."

Monk Gibbon, SPECTATOR

MOUNTAINS AND MOORLANDS—*W. H. Pearsall*

" An integrated study of upland Britain that will be accepted as a classic and which only a man of Professor Pearsall's exceptional breadth of vision could have accomplished." YORKSHIRE POST

" Professor Pearsall's work is a model of contruction, for not only does he deal with the making of the mountains and the moors, but he shows how they alone have been responsible for sustaining very definite types of flora and fauna." THE FIELD

" It is doubtful whether any other author could, single-handed, have produced such a well-balanced picture of the wild life of any area . . . his book is illustrated by a really magnificent series of colour photographs of hill country." WEEKLY SCOTSMAN

MUSHROOMS AND TOADSTOOLS—*John Ramsbottom*

" The excellent New Naturalist series . . . is comprehensive and scholarly and beautifully illustrated. The general reader may consult it to discover which fungi should not be eaten, to learn the true facts about dry rot, or the amazing story of the development of penicillin. Students will consult it on every aspect of mycology." JOHN O'LONDONS

" This supplies a long-felt want . . . exhaustive and authoritative book . . . a scientific work by one who is probably our greatest authority on the subject . . . an immensely interesting book. The illustrations in *Mushrooms and Toadstools* are exceptionally good. Many of those from colour photographs are quite outstandingly beautiful." ILLUSTRATED LONDON NEWS

BRITISH PLANT LIFE—*W. B. Turrill*

" We strongly recommend it to all academic students of botany and to many more amateur specialists who are making flowering plant life their special line." *L. J. F. Brimble*, NATURE